The Natural History of
THE LONG EXPEDITION TO THE
ROCKY MOUNTAINS
1819–1820

Ottoes. may 1820)

The Natural History of
THE LONG EXPEDITION TO THE
ROCKY MOUNTAINS
1819–1820

Howard Ensign Evans

New York   Oxford
OXFORD UNIVERSITY PRESS
1997

Oxford University Press

Oxford   New York
Athens   Auckland   Bangkok   Bogotá   Bombay
Buenos Aires   Calcutta   Cape Town   Dar es Salaam
Delhi   Florence   Hong Kong   Istanbul   Karachi
Kuala Lumpur   Madras   Madrid   Melbourne
Mexico City   Nairobi   Paris   Singapore
Taipei   Tokyo   Toronto

and associated companies in
Berlin   Ibadan

Library of Congress Cataloging-in-Publication Data
Evans, Howard Ensign.
The natural history of the Long Expedition to the Rocky Mountains
(1819–1820) / by Howard Ensign Evans.
p. cm.   Includes bibliographical references and index.
ISBN 0-19-511184-2—ISBN 0-19-511185-0 (pbk.)
1. Natural history—West (U.S.)
2. West—Description and travel.
3. Stephen H. Long Expedition to
the Rocky Mountains (1819–1820)   I. Title.
QH104.5.W4E93 1997   96-26655   508.78—dc20

1 3 5 7 9 8 6 4 2

Printed in the United States of America
on acid-free paper

Contents

PREFACE

DRIVING ALONG INTERSTATE 25 FROM LONGMONT to Pueblo, Colorado, one scarcely notices the landscape for the innumerable signs, advertising motels, restaurants, airlines, and automobiles. Traffic floods the highway, especially in Denver, Colorado Springs, and Pueblo: people rushing to and from work; trucks carrying furniture, beer, foodstuffs, or whatever; recreational vehicles maneuvering to an exit that has a gasoline station. Always the mountains of the Front Range to the west, suggesting places less noisy and smog-ridden, where deer run and trout splash. But even they are criss-crossed with roads and splotched with houses. The radio tells of new industry, new jobs, new shopping malls, and in the next breath, unabashedly, of more pollution, more water problems. It is a world addicted to the growth of the human enterprise. Yet a little more than 170 years ago—hardly a moment in the clock of history—the land was empty of all but Indians and the plants and game on which they subsisted.

In 1820, twenty-two men—military personnel and "scientific gentlemen"—struggled along the Front Range, living off the land, recording rivers and landforms, shooting birds, plucking plants, and catching lizards and insects to preserve for study. They were often thirsty and hungry, sometimes ill, and always tired. But theirs was an experience awarded to a chosen few, that of seeing and recording for the first time a land never before visited by persons trained in European scientific traditions. This was the Long Expedition, a small party with a tight time schedule, launched with unrealistic goals and inadequate financial support by a government only reluctantly coming to terms with the vast new lands that Thomas Jefferson had acquired in 1803.

Their story has been told before, but without due recognition of the contributions of the expedition's naturalists, particularly botanist Edwin James and zoologist Thomas Say. They were the first to provide scientifically acceptable names and descriptions of plants and animals of the High Plains and the Front Range of the Rocky Mountains. In a sense they are still there, whenever a waxflower (*Jamesia*) blooms from a rocky crevice or a Say's phoebe (*Sayornis saya*) snags a fly from its perch on a fence post.

To best appreciate the contributions to natural history made by the expedition, it is necessary to strip away (as I have done) many details of logistics, of topography and geology, and of experiences with the Native Americans that make up so much of the text of the original *Account of an Expedition from Pittsburgh to the Rocky Mountains*, as prepared by Edwin James in 1823. Those wishing to consult the original report will find it in four volumes of Reuben Gold Thwaites's *Early Western Travels, 1748–1846*, and in two volumes of Readex Microprint. They are available in many libraries. More recently, Maxine Benson edited James's *Account* in one volume, *From Pittsburgh to the Rocky Mountains*, briefly summarizing some sections of the narrative and leaving out the footnotes (many of which contain natural history observations). Benson's volume includes many of the sketches and paintings by Titian Peale (assistant naturalist) and Samuel Seymour (landscape artist); some are in color, and most were not included in the original report of the expedition.

James's *Account* was "compiled from the notes of Major Long,

Mr. T. Say, and other gentlemen of the party," as acknowledged on the title page (James himself joined the expedition only during its second year). James made much use of his own unpublished diary and of Titian Peale's diary, only part of which survived, eventually to be published by A. O. Weese as "Journal of Titian Ramsey Peale, Pioneer Naturalist." Thomas Say contributed in important ways to the *Account* and made many references in his later publications to sites where specimens had been collected. The *Account* also includes extracts from Stephen H. Long's report to Secretary of War John C. Calhoun as well as appendixes of astronomical and meteorologic records and vocabularies of Indian languages.

Actually, Captain Thomas Biddle was the expedition's official journalist during the first year, but he failed to keep the necessary information and left the expedition after a few months. He was replaced by Captain John R. Bell, whose informative journal was not used by James in compiling the *Account*. Bell was to have submitted his report to Secretary of War Calhoun, but it evidently never reached Calhoun's office. Bell died in upstate New York only five years after the expedition's return, and his handwritten manuscript came into the possession of a family that later moved to California. There it was discovered in 1932 by Harlin M. Fuller, who, with co-editor LeRoy R. Hafen, published *The Journal of Captain John R. Bell.*

An excellent review of this and other expeditions led by Stephen Long is presented in Roger L. Nichols and Patrick L. Halley's book *Stephen Long and American Frontier Exploration.* Most recently, in *Retracing Major Stephen H. Long's 1820 Expedition,* George J. Goodman and Cheryl A. Lawson retrace Long's route from the Missouri to Fort Smith, Arkansas, during 1820. They precisely identified many of the expedition's campsites and visited many of them themselves. A major part of their book contains a species-by-species listing of the several hundred plants collected by the expedition's botanist, Edwin James, noting the locality in which each was probably collected. Their book does much to fill in details of an important chapter in western history and is a tribute to the botanical accomplishments of Edwin James.

I have often let the expedition's participants speak for themselves, even though their language is sometimes quaint and their spelling and grammar are not always "by the book." (Bell's journal was never planned for publication, and it is less literate than James's *Account.*) Rather than using footnotes, I have indicated who is speaking by the use of their initials: JB, John Bell; EJ, Edwin James; SL, Stephen Long; TP, Titian Peale; TS, Thomas Say. Quotations from James are from the *Account* rather than from his diary, except as noted. It must be borne in mind that since James did not join the expedition until the second year, all quotes from him during the first year are based on information he received from Say, Peale, Augustus Jessup, and William Baldwin.

During the first summer, traveling on a recalcitrant steamboat through country already thinly settled, the expedition got only as far as the vicinity of present-day Omaha, Nebraska. This part of the trip I shall cover more briefly than the much more adventurous and productive second summer, which took the exploratory party by foot and horseback to the Rockies and back.

By interweaving the impressions of the expedition members, I have tried to provide an opportunity for moderns who admire the out-of-doors to visit vicariously lands that have been vastly transformed. Long felt that the semiarid country to the west of the hundredth meridian would prove to be a barrier to settlement of the West. Modern technology has proved him wrong. Long has been condemned for his misjudgment of the land's possibilities and for his failure to fulfill several unrealistic goals that had been set for him. But there were major accomplishments: for the first time the broad area between the northern route of Lewis and Clark and the more southerly route of Pike was mapped with reasonable accuracy; and for the first time, naturalists were able to report on the rich and often surprising fauna and flora of the central western plains and the Front Range of the Rockies. More could hardly be asked of a small band

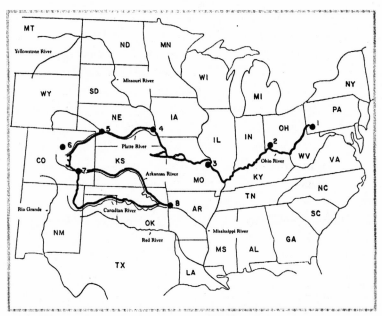

Route of the Long Expedition, with major landmarks indicated: (1) Pittsburgh, May 3, 1819; (2) Cincinnati, May 9–18, 1819; (3) St. Louis, June 9–12, 1819; (4) Engineer Cantonment, September 17, 1819 –June 6, 1820; (5) crossing the Platte, June 22, 1820; (6) Long's Peak, sighted June 30, 1820; (7) crossing the Arkansas and separation of the two parties, July 22–24, 1820; (8) arrival at Fort Smith, September 9–13, 1820.

of inexperienced and ill-equipped explorers as they plodded across landscapes that had rarely been visited by Americans of European origin. History books have consistently ignored the Long Expedition or denigrated its accomplishments. It is an episode in history that deserves to be remembered and re-evaluated.

I am particularly indebted to two western history enthusiasts who read earlier versions of the manuscript and made many helpful suggestions: Richard G. Beidleman, formerly of Colorado College, Col-

PREFACE

orado Springs; and Robert E. Heapes, of Parker, Colorado. Kenneth Haltman, of Michigan State University, has also been most helpful. Mary Alice Evans has borne with me through the book's long gestation, always at hand when help was needed and a companion as we visited many sites along the route of the expedition.

For assistance in obtaining prints of sketches and paintings made by the expedition's artists, I am indebted to Richard S. Field and Bernard Noveloso, Yale University Art Gallery; Larry K. Mensching, Joslyn Art Museum, Omaha; Carol M. Spawn, Academy of Natural Sciences of Philadelphia; Beth Carroll-Horrocks and Miriam B. Spectre, American Philosophical Society; Thomas V. Lange, The Huntington Library, San Marino, California; and Karen L. Otis, Museum of Fine Arts, Boston.

Fort Collins, Colo.   H. E. E.
September 1996

The Natural History of
THE LONG EXPEDITION TO THE
ROCKY MOUNTAINS
1819–1820

One

SETTING THE STAGE

THE LOUISIANA PURCHASE OF 1803 APPROXIMATELY doubled the size of the United States, adding 800,000 square miles of ignorance—land that had never been well explored or adequately mapped. Even the northern and southern boundaries, with British Canada and Spanish Mexico, had not been explored. To the west, Louisiana ended in ranges of mountains of unknown dimensions. That such mountains were there had been shown by Father Silvestre Vélez de Escalante, who in 1776 had crossed from Santa Fe to what is now northeastern Utah, seeking in vain for a convenient route to California, and by Alexander Mackenzie, who had crossed the Canadian Rockies to the Pacific in 1793. Fur trappers and traders—most of them French or Spanish and often illiterate—had also penetrated parts of the mountains, bringing back tales of white bears and wild Indians.

By 1803, even before the treaty with France had been formalized, Thomas Jefferson had persuaded Congress to appropriate $2,500 to outfit an expedition that he hoped would cross the mountains "even to the Western Ocean," and Meriwether Lewis and William Clark were assembling supplies and equipment for their Corps of Discovery. Following their return, after two years and four months in the wilderness, Americans were finally to learn of the vastness of their western lands and much about its geographic features, its soils, and its inhabitants.

The Lewis and Clark Expedition included a core of twenty-nine men, along with several others who helped move their boats up the Missouri to Fort Mandan, in present-day North Dakota. There they spent the first winter and engaged the services of Touissant Charbonneau and his wife, Sacagawea, who soon captured the imagination of the nation. From Fort Mandan, their route took them across what is now Montana and Idaho, and then down the Columbia to its mouth. There, building a post they called Fort Clatsop, they spent the second winter, and then returned over roughly the same route. The story of Lewis and Clark's exploits has been told many times and needs no retelling here. The expedition opened the Northwest for American fur trappers and traders; already by 1807, Manuel Lisa had built a trading post at the junction of the Yellowstone and Bighorn Rivers, deep in what is now Montana.

Quite different was the expedition of Zebulon Pike, initiated in July 1806, even before the Corps of Discovery had returned. Pike's instructions came not from Washington, but from General of the Army James Wilkinson, governor of Louisiana Territory, a man who had dreams of establishing a separate empire in the western plains and the Southwest. His orders to Pike were to pacify the Pawnees and the Comanches, and then to explore the sources of the Red River, the supposed boundary between American and Spanish lands. Having accomplished the first mission, after a fashion, Pike found himself on the upper Arkansas River at the threshold of the Rockies—and it was already mid-November. Despite the bitter weather, Pike and his small band of sixteen penetrated the mountains as far as South Park (near modern-day Fairplay, Colorado) and nearly to Leadville. He was close to the sources of both the Arkansas and South Platte Rivers. Another, north-flowing stream he believed to be the source of the Yellowstone, a major branch of the Missouri, and his map showed his trail meeting the source of that river: "I have no hesitation [he wrote] in asserting that I can take a position in the mountains, whence I can visit the source of any of those rivers in one day." The rivers he cited were the Yellowstone, the Platte, the Colorado, the Arkansas, and the Rio Grande! (William Clark, too, had mapped the origin of the Yellowstone not far from that of the Rio Grande and the Colorado.)

After enduring many hardships, Pike and his men retreated to the eastern foothills of the Rockies, where they built a blockhouse for shelter. But there was little game to be found, and Pike left with a smaller group to cross the Sangre de Cristos. There he found another river, which might have been the river he had been commissioned to find: the Red. (It was, of course, the Rio Grande.) After building another stockade, he was captured by the Spanish and taken to Santa Fe and later to Chihuahua, where he was released several months later. It is probable that Wilkinson had planned for this to happen, as he wanted Pike to report on Spanish settlements and military forces. Pike's report, published in 1810, has been described by one biographer as "poorly organized, unreliable . . . scientifically and geographically incorrect, and in many places dishonest." By his own admission, Pike had no qualifications as a naturalist, and he lacked the "time and placidity of mind" required to study the plants and animals he encountered. Nevertheless Pike's account of the rich villages of the Southwest helped to set the stage for the development of the Santa Fe Trail.

While in South Park, Pike was impressed by the size of a recent encampment, and in Santa Fe he met a trader, James Purcell (or Pursley, as he called him), who explained that he had camped there in 1805 with a large group of Kiowa Indians. This was a year and a half before Pike reached the area, so it seems fair to credit Purcell with being the first American of European descent to discover the Front Ranges of Colorado. Purcell told Pike that he "had found gold in the head of La Platte, and had carried some of the virgin mineral in his shot-pouch for months; but that, being in doubt whether he should ever again behold the civilized world . . . he threw the sample away." Pike spoke of Purcell as "a man of strong natural sense and dauntless intrepidity."

Purcell was a Kentuckian who had hunted and trapped in Louisiana Territory since 1799. In 1805 he had been hired by a trader to make contact with the Kiowa Indians. But the Kiowas were being driven south and west by the Sioux, and along with them went Purcell and several companions. The camp in South Park was said to have contained as many as 2,000 Indians and more than 10,000 horses. The Indians sent Purcell to Santa Fe to establish trade relations with the Spanish, but he remained there, later becoming a Mexican citizen and settling in Sonora.

The supposition of both Clark and Pike that the Yellowstone River arose in central Colorado demonstrates the contemporary lack of appreciation of the vast area between the watershed of the Missouri and that of the Arkansas. It was not until 1811 that a group of traders made the trek from the Missouri to the Arkansas (it took them "forty or fifty" days) and not until 1816 that their trip was made known, and under odd circumstances, in a letter to the editor of a newspaper, the *Missouri Gazette*. The letter was written by Ezekiel Williams and was in response to an article concerning the discovery of a grave in central Missouri believed to be that of Jean Baptiste Champlain, who, along with Williams and several others, had been sent by Manuel Lisa from his trading post on the Missouri toward Santa Fe, in the hope of establishing trade with the Spanish. Several in the party perished, but Williams and Champlain survived to procure a valuable load of furs. But, said the report, when nearly back to Boone's Lick, Missouri, Williams "coolly and premeditatedly committed one of the most inhuman and outrageous acts of cruelty that the annals of history can produce, by putting to death the friend of his bosom for the sake of lucre!" The source of this story was not stated, except that it was "from a gentleman of respectability."

All of this Williams emphatically denied, explaining that he had left Champlain with the Arapahos when he returned via the

Arkansas River, and on a later trip found that the Arapahos had killed him. This story was confirmed by Robert Stuart, leader of John Jacob Astor's American Fur Company, who learned from the Shoshones that Champlain had been killed by the Arapahos in 1812. Members of the Long Expedition were to hear of the discovery of the body as they crossed Missouri in 1819. If the body was that of Champlain (as seems doubtful), it remains to be explained how it got from Colorado to Missouri.

During the late 1820s, Williams served as a guide on the Santa Fe Trail. Later "Old Zeke" settled on a farm in Benton County, Missouri, where for a time he served as judge and postmaster. Although his name has nearly been forgotten, it was Williams who first pointed out that there was a vast, barely explored region between the Missouri and Arkansas Rivers. Whether his letter to the *Missouri Gazette* describing his experiences was known to the eastern establishment of the time is a moot point.

Once Manuel Lisa and others of his Missouri Fur Company had learned how to run the gauntlet of the Sioux and other tribes along the Missouri, that river became the scene of much activity. Between 1807 and 1820, Lisa and his companions went up and down the river several times. They traveled in keelboats poled or rowed through the turbulent waters or pulled upstream from the banks by ropes; only now and then could they use wind power to replace muscle power. Trappers and traders were dispatched from posts along the river, John Colter traveling even as far as the Wind River Ranges, Jackson Hole, and Yellowstone. Colter, like several others in Lisa's employment, was a veteran of the Lewis and Clark Expedition.

In 1811, Astor's Pacific Fur Company dispatched Wilson Price Hunt and his party of "Astorians" up the Missouri in four keelboats. Later they would trek, with much difficulty, all the way to the Pacific, and the following year Robert Stuart would lead the return trip east. With Hunt as he ascended the river were two notable botanists, Thomas Nuttall and John Bradbury. Bradbury would write of his adventures in his book *Travels in the Interior of America in the*

Years 1809, 1810, and 1811 (1817). With Lisa, that same year, was geographer Henry Marie Brackenridge, who wrote of his travels in *Views of Lousiana: The Journal of a Voyage up the Missouri River, in 1811* (1814). So the Missouri River, at least as far as the Dakotas, was reasonably well known compared with most parts of the West.

One would have thought that the next government-sponsored expedition might have had as its major goal the exploration of some of the vast areas to the south of the Missouri that were still little, if at all, visited, but that is not the case. Following the War of 1812, John C. Calhoun, Secretary of War under President James Monroe, was concerned with British influence in the northern part of Louisiana Territory and the danger from Indians who had allied themselves with the British. Calhoun was then thirty-seven, and had been in the cabinet for two years. This was long before he became an advocate of states' rights; he was then an ardent nationalist and expansionist. "Distance and difficulties are less to us than any people on earth," he had remarked in a speech to Congress. There was genuine fear of a third war with England, working through the Indians to block expansion to the west.

Calhoun proposed sending troops up the Missouri to establish a fort at the mouth of the Yellowstone River (near the present boundary between North Dakota and Montana). There they would impress the Indians with the power of the Americans and at the same time serve notice to Canadian trappers and traders that some of their favorite beaver country was now part of the United States. President Monroe supported the plan fully, as he wrote to his Secretary of War.

> The people. . . . look upon it as a measure better calculated to preserve the peace of the frontier, to secure to us the fur trade and to break up the intercourse between the British traders and the Indians, than any other which has

been taken by the government. I take myself very great
interest in the success of the expedition, and am willing
to take great responsibility to ensure it.

The press was no less enthusiastic about the Yellowstone Expedition,
as it came to be called. The *Missouri Gazette* reported that "the plan
has attracted the attention of the whole nation, and there is no
measure which has been adopted by the present administration that
has received such universal commendation." That the expedition
was to make use of the recently invented steam engine especially
fired the imaginations of many. In a letter to a newspaper, one cor-
respondent surmised that it would lead to "safe and easy communication to China [and] ten years shall not pass away before we shall
have the rich productions of that country transported from Canton
to the Columbia, up that river to the mountains, over the mountains
and down the Missouri and Mississippi, all the way (mountains and
all) by the potent power of steam."

Both steamboats and steam railways were novel in the early
nineteenth century, and the public was as much excited about them
as we are about space probes. Although Colonel Henry Atkinson,
who had been placed in charge of the military arm of the expedition,
was well aware that human- and wind-powered keelboats had successfully plied the Missouri, they would not do for so grand an expedition. A contract was made with Colonel James Johnson to build
five steamboats, to be named the *Jefferson*, the *Calhoun*, the *Johnson*,
the *Exchange*, and the *Expedition*. The steamboats were hastily con-
structed and proved to be more expensive than had been calculated
and much less effective than had been hoped in traversing the Ohio
and the Missouri. The plan was to transport nearly 1,000 soldiers
and their equipment to the mouth of the Yellowstone. It was assumed that the War Department would save money in the process,
since the troops could live mainly on the abundant game. As we
shall see, none of the ships came anywhere near reaching their

planned destination, and many of the troops succumbed to scurvy during the winter of 1819/1820.

From the beginning, it was planned to add an exploratory arm to the expedition, to be commanded by Major Stephen H. Long. That the enterprise eventually came to be called the Long Expedition rather than the Yellowstone Expedition reflects the relative success of the two arms. Long had only just returned from expeditions on the upper Mississippi River and in Arkansas. He was given a more or less free hand to design his own steamboat and to select his personnel. His steamboat, the *Western Engineer,* had a shallower draft than any of the five ships of the military contingent, and may have been the first stern-wheeler ever built. It proved far better suited for river travel than did Colonel Johnson's boats, but its operation was not without problems.

As a school principal in Germantown, Pennsylvania, before joining the military, Long had become acquainted with members of the American Philosophical Society in nearby Philadelphia, then the intellectual hub of the country. The idea of asking several members of the society to accompany him on this new venture was his, but Calhoun was easily persuaded. Several were eager to do so despite the fact that they were offered little salary and would have to supply some of their own equipment.

The expedition left Pittsburgh in May 1819. Calhoun's orders to Long were (in part) as follows:

> You will first explore the Missouri and its principal branches, and then, in succession, Red River, Arkansa [sic] and Mississippi, above the mouth of the Missouri.
>
> The object of the Expedition, is to acquire as thorough and accurate knowledge as may be practicable, of a portion of our country, which is daily becoming more interesting, but which is as yet imperfectly known. With this view, you will permit nothing worthy of notice, to escape

your attention. You will ascertain the latitude and longitude of remarkable points with all possible precision. You
will if practicable, ascertain some point in the 49th parallel of latitude, which separates our possessions from
those of Great Britain. A knowledge of the extent of our
limits will tend to prevent collision between our traders
and theirs.

You will enter in your journal, everything interesting
in relation to soil, face of the country, water courses and
productions, whether animal, vegetable, or mineral.

You will conciliate the Indians by kindness and presents, and will ascertain, as far as practicable, the number
and character of the various tribes, with the extent of
country claimed by each.

Great confidence is reposed in the acquirements and
zeal of the citizens who will accompany the Expedition
for scientific purposes, and a confident hope is entertained,
that their duties will be performed in such a manner, as
to add both to their own reputation and that of our country.

To his orders, Calhoun appended a copy of Jefferson's instructions
to Meriwether Lewis, hoping that they might provide "many valuable suggestions."

That Long was asked to explore the boundary with Canada and
that with Mexico (the Red River) in one expedition—along with
the Mississippi, the Missouri and its tributaries, and the Arkansas—
reveals the prevailing ignorance of the vastness of the western lands.

Calhoun and Long placed much emphasis on the importance
of the scientific personnel who would accompany the expedition. It
is sometimes said that this was the first expedition to the West that
included trained naturalists. This statement requires qualification.
Lewis and Clark made very substantial observations on natural his-

tory along their route to the Pacific and back. Jefferson had seen to it that Lewis had a "crash course" in natural history under Benjamin Smith Barton and other Philadelphia intellectuals, and Clark proved to be a superb observer and geographer despite his limited education. Their journals contain a wealth of novel information. In his book *Lewis and Clark: Pioneer Naturalists*, Paul Russell Cutright includes a list of the biological discoveries of the expedition. It requires forty-seven pages!

Here we need to distinguish between the act of discovery and the documentation of a plant or an animal in the scientific literature. To be formally established, a plant or an animal must be given a Latinized double name (genus and species), following rules established by Swedish naturalist Carl Linnaeus in the mid-eighteenth century. The name must be followed by a detailed description and a statement of the locality in which the plant or animal was collected. The specimen should then be deposited in a reputable institution where it will be preserved and can be studied by others.

Lewis and Clark did collect both plants and animals and tried to see to it that they reached authorities in the East. Lewis's descriptions were often so precise that many of the species he discussed can now be identified. However, these explorers shied away from using Linnaean nomenclature and left that task to others. Thus Clark's nutcracker and Lewis's woodpecker were described and illustrated by pioneer ornithologist Alexander Wilson in 1811, and two of the more spectacular plants they discovered were named *Lewisia* and *Clarkia* by botanist Frederick Pursh in 1814. Philadelphia zoologist George Ord described the grizzly bear, the pronghorn, and several other mammals sent back by Lewis and Clark. The journals of these explorers were printed in 1814, after the untimely death of Lewis, but only after the editor, Philadelphia lawyer Nicholas Biddle, had improved their spelling and greatly condensed their narrative. A second volume, to have covered the scientific accomplishments of the expedition, was never published because of the illness of its editor, Benjamin Smith Barton. In 1893, Elliott Coues resurrected the

Biddle edition, edited it extensively, and added copious notes derived from a careful reading of the eighteen volumes of original manuscript. The first printing in full of the original journals of Lewis and Clark, with much additional material, was edited by Reuben Gold Thwaites and published in eight volumes in 1904 and 1905 (reprinted in 1969 by Arno Press, New York). Still more recently, from 1986 to 1990, the University of Nebraska Press published the journals in seven volumes, edited by Gary Moulton, making use of still additional material and the techniques of modern historical research. So it has taken nearly two centuries to appreciate fully the accomplishments of the Lewis and Clark Expedition.

The naturalists of the Long Expedition were better trained as systematists and did not hesitate to provide formal names and descriptions of many of the plants, animals, and geologic formations they encountered. Only in this sense were their observations more "scientific" than those of Lewis and Clark. Since the report of the Long Expedition was published in full in 1823, the discoveries of its naturalists became well known long before those of Lewis and Clark. The report was an important contribution to science of the day, particularly the account of the second year, when Long and his men left the Missouri to explore the western plains and the Front Range of the Rockies.

The first year of Long's expedition, spent on the Ohio and the Missouri, was by no means in territory virgin to naturalists. Botanists Bradbury and Nuttall had preceded the expedition up the Missouri, and Nuttall had explored the lower Arkansas basin in 1819; both published accounts of their trips. A bowdlerized edition of Lewis and Clark's journals had been published in 1814. In 1808, Alexander Wilson published the first part of his *American Ornithology*, in which he formally described some of the birds collected by Lewis and Clark, illustrating them with colored plates. In 1814, Frederick Pursh published his *Flora Americae Septentrionalis*, in which he named and described many of the plants collected by Lewis and Clark, and (without authority) some of those collected by Bradbury and by Nuttall. Pursh, once called "one of the most active and apparently un-scrupulous early Philadelphia botanists," had never been west, but his *Flora* proved valuable as a preliminary guide to some of the western plants. In 1818, Nuttall published *The Genera of North American Plants and Catalogue of the Species to the Year 1817*, based in part on his experiences on the Missouri in 1811.

It is not clear how much of the available literature was carried by the Long Expedition. It is known that the explorers carried the maps made by Clark and by Pike, as well as the Lewis and Clark *Journals* and Alexander von Humboldt's *Personal Narrative of Travels to the Equinoctial Regions of America*. Whether they carried Ord's descriptions of some of the mammals collected by Lewis and Clark, Wilson's *American Ornithology*, Pursh's *Flora*, or Nuttall's *Genera* is uncertain, but in any case these were available to the naturalists as they prepared the report of the expedition for publication.

To understand the contributions of the expedition's naturalists, it is necessary to appreciate the rather primitive state of natural history in their time. Linnaeus had been dead for only forty years, and Darwin's *On the Origin of Species* was forty years in the future. Knowledge of the natural history of the eastern states was still sketchy, and the western lands were a vast unknown. That the expedition's naturalists made mistakes and provided descriptions of natural objects that are inadequate by modern standards is understandable, particularly when we consider the difficulties under which they were working.

The naturalists used the Linnaean system of nomenclature, but in their time some procedures were less well established than they are today. Nowadays generic names are capitalized and species names uncapitalized, and both are placed in italics to set them off from the text; but in the 1820s these rules were not always followed. The name of the person who described the species is often placed after the name, sometimes abbreviated. The naturalists did this irregularly ("N" following a plant name, for example, refers to Nuttall). When a genus name is repeated, it may be abbreviated by using only the first initial (for example, C. *latrans*, when the genus *Canis* has been spelled out just previously). The naturalists sometimes described as new species ones that had actually been described earlier by someone else. In this case the "law of priority" prevails: the earlier name is accepted. For example, Thomas Say described the mule deer, naming it *Cervus macrotis*. He was unaware that Constantine Rafinesque had described it just a few years earlier, calling it C. *hemionus*, the species name that is now accepted.

Pioneer naturalists tended to use broad, all-embracing genera. As science advanced, more species became known and their relationships better understood. This has led biologists to divide the old, inclusive genera into several genera of more precise definition. It is no discredit to the expedition's naturalists that their species have now often been placed in different genera than those in which they placed them. Science progresses by improving the superstructure as knowledge is added.

The species of plants and animals newly described from specimens collected by the expedition I have listed in the appendixes. The lists, although not complete, include slightly over 300 species, a considerable accomplishment for a small, ill-equipped group that moved rapidly through rough and mostly unexplored country. It was not always easy for the naturalists to collect and prepare specimens and write up accompanying notes when they were traveling by foot and horseback twenty or more miles a day, often through heat, storms, and biting insects. Of course, the discovery of new species was not the be-all and end-all of the expedition. The naturalists learned much about the distribution of plants and animals and about their living conditions. They also recorded data about the rocks and landforms and made an effort to learn as much as possible about the Native Americans they met—though much of that information is omitted from this book. Long and his lieutenants regularly determined their longitude and latitude, and from these data they prepared maps that were a great improvement over any then available. The expedition also served notice to the Native Americans and to the Spanish and British that this land was now part of the United States, something that was clearly on the minds of President Monroe and Secretary of War Calhoun when the trip was authorized.

The final report of the expedition, published in both Philadelphia and London in 1823, had a wide readership, but it is safe to say that most readers were more interested in the descriptions of landscapes and Indians than in those of newly discovered plants and animals. Within a few years, the findings of Long and his men had become incorporated into Easterners' perceptions of the West, provoking thoughts of escape to virgin lands and the background for many a novel about the mythic West.

Unlike the journals of Pike and of Lewis and Clark, the *Account* of the Long Expedition was illustrated. The American edition contained eight engravings made from paintings done by the expedition's artists, while the English edition included four additional plates, two of them hand-colored landscapes. For the first time, views of the western plains and of the slopes of the Rockies, along with their native inhabitants, were available to the public. Long's was the first of several expeditions to the West that documented its travels by the work of artists. Not until the 1860s was photography available and widely used by travelers through the West.

The two artists of the expedition actually made many more sketches than appeared in the *Account*. According to Kenneth Haltman, who has made a special study of the expedition's artistic heritage, Samuel Seymour produced about 150 sketches or paintings; Titian Peale, as many as 235. Many of Peale's have survived, and some have appeared in subsequent publications by diverse authors. Sketches made in the field often were later used to make more finished paintings after the men had returned to Philadelphia. Thus an extensive visual record of the expedition became available to scholars.

$\mathcal{T}wo$
CAST OF CHARACTERS

IN THE SPRING OF 1819, TWENTY-FOUR men gathered in Pittsburgh with their personal effects, ready to depart on Major Stephen H. Long's steamboat, the *Western Engineer*. The military contingent had preceded them in five steamboats. Long's party included a small group of army personnel as well as five "scientific gentlemen" and a crew of six:

MILITARY
Major Stephen H. Long	Commander
Major Thomas Biddle, Jr.	Journalist
Lieutenant James D. Graham	Assistant topographer
Cadet William H. Swift	Second assistant topographer
Sergeant Samuel Roan	
Eight privates	

SCIENTIFIC
Dr. William Baldwin	Botanist and surgeon
Thomas Say	Zoologist
Titian R. Peale	Assistant naturalist

Augustus Jessup	Geologist
Samuel Seymour	Artist

CREW

Benjamin Edwards	Steamboat engineer
Thomas Boggs	Pilot
Isaac Kimball	Carpenter
L. R. Kinney	Clerk
Two "boys"	

Major Benjamin O'Fallon, Indian agent for the tribes along the Missouri River, became attached to the expedition informally at a later date, as did his assistant and interpreter, John Dougherty. The crew and the soldiers are rarely mentioned in the narrative of the expedition, though doubtless all were important in bringing the *Western Engineer* to its destination in the fall of 1819. As to the officers and scientists, a bit more needs to be said.

Stephen Harriman Long came from a New Hampshire farming family. At Dartmouth College, he was a leader among the students and was elected to Phi Beta Kappa, the national honorary fraternity. After a year of teaching school in New Hampshire, he accepted a position as a school principal in Germantown, Pennsylvania. There he became acquainted with members of the American Philosophical Society in nearby Philadelphia. In his spare time, he did some surveying and acquired a reputation as an inventor of machinery. This brought him to the attention of officers of the army's Corps of Topographical Engineers, and in 1814 he was commissioned as a second lieutenant. After a year of teaching mathematics at West Point, he was sent west to inspect forts and gather information on the streams and soils of Illinois and the upper Mississippi basin. He enjoyed traveling in the wilderness and was enthusiastic about the landscape and the potential of the prairies for agriculture.

In 1817 Long was sent to Arkansas, where the Cherokee and

Osage Indians had been waging periodic wars. There he explored parts of the lower Arkansas and Red Rivers and was instrumental in establishing Fort Smith, close to the present Arkansas–Oklahoma border. As before, he traveled primarily by water, using human- and wind-powered skiffs, though as an engineer he was weighing the feasibility of using the recently invented steamboat on western rivers and had even written to President Monroe concerning this possibility. Long's reports were well received in Washington, and many of his recommendations concerning forts and lines of communication were implemented over the next few years. One of Long's chief characteristics as an explorer became evident during these expeditions: he was forever eager to push forward, often leaving insufficient time to observe the country in much detail. Nevertheless, with his experience in the wilderness and his contacts in the army and among eastern intellectuals, he was ideally suited for his new assignment. In 1819 he was thirty-five years old and had attained the rank of major.

Although steamboats had been in use for several years, none had ever attempted the ascent of the turbulent and unpredictable Missouri. Such was Long's reputation as an engineer that Secretary of War Calhoun gave him permission to design a steamboat for his forthcoming expedition and to oversee its construction. The craft he built was seventy-five feet long and only thirteen feet wide, with a paddle wheel in the rear, features designed to take advantage of narrow river channels without being delayed by logs and debris in the stream. Although the ship's performance fell short of expectations, it was able to ascend the Missouri farther than any of the five that Johnson had built for the military contingent.

Thomas Biddle, Jr., came from a prominent Philadelphia family. He fought in the War of 1812 with distinction and had asked to join the expedition. Long assigned him the duty "to record all transactions of the party that concern the objects of the expedition, to describe the manners and customs, etc., of the country through which we may pass; to trace in a compendious manner the history of the towns, villages, and tribes of Indians" and so forth. There is no evidence that he did any of these things. He and Long quarreled from the start, and he left the expedition after only three months, joining the staff of Colonel Henry Atkinson. According to Roger Nichols and Patrick Halley, he remained in the army and "died in 1831 during a duel with Congressman Spencer Petis on Bloody Island near St. Louis."

James Duncan Graham was a Virginian who had graduated from West Point in 1817; in 1819 he was twenty years old and a first lieutenant in the artillery. William Henry Swift was brought up in Massachusetts; he was only nineteen when he embarked on the *Western Engineer*. He did not officially graduate from West Point until July 1819; he was commissioned as a second lieutenant while well on his way west. Graham and Swift were Long's chief assistants in a "prime objective of the expedition," making "a topographical description of the country to be explored." They were also expected to "attend to drilling the boat's crew, in the exercise of the musket, the field-piece, and the sabre."

To be botanist of the expedition, Long chose Dr. William Baldwin, a physician with a strong interest in botany. A Philadelphia Quaker, Baldwin had been trained in medicine at the University of Pennsylvania. As a young man, he showed symptoms of tuberculosis, and in the effort to improve his health he enlisted as a surgeon on a merchant ship bound for China. On his return, he settled in Georgia, where he traveled about on foot collecting plants when not occupied as a physician. In 1817 he sailed for South America on the frigate *Congress*, giving him the opportunity to explore another continent. After he returned his health remained poor, and he hoped to recover by joining an expedition to the West. After reviewing his application, Long felt that Baldwin "stands alone" among botanists. In fact, Thomas Nuttall—younger than Baldwin, in better health, and already the author of a significant botanical treatise—was eager to join the expedition. But Long required a botanist who could double as a physician, and Nuttall had no medical training. Long's expectations of his botanist were high.

A description of all the products of vegetation, common or peculiar to the countries we may traverse, will be required of him, also the diseases prevailing among the inhabitants, whether civilized or savage, and their probable causes, will be subjects for his investigations; any variety in the anatomy of the human frame, or any other phenomena observable in our species, will be particularly noted by him. Dr. Baldwin will also officiate as physician and surgeon for the expedition.

When he joined the expedition, Baldwin was forty-one. After he had made the appointment, Long learned that a younger man, Dr. John Torrey, had expressed an interest in the expedition. He offered Torrey the position of geologist, but Torrey declined when Long could not promise him "whether a pecuniary consideration shall be allowed." (It was later decided to pay the scientific personnel $2.20 a day.) As luck would have it, Baldwin died a few weeks into the expedition, while Torrey went on to become the major American botanist of his time. Torrey eventually described many of the plants collected in the West by Edwin James, who replaced Baldwin in 1820.

Long's choice as zoologist was another Philadelphian and member of the American Philosophical Society, Thomas Say. Although sometimes referred to as "Dr. Say," he had little formal education. A great-grandson of pioneer naturalist John Bartram, Say failed as a pharmacist but at an early age began collecting natural history specimens. He came to haunt Peale's Museum in Philadelphia, where he often worked late into the night and sometimes slept beneath the mastodon skeleton—the only space clear enough to accommodate him. Say was a charter member of the Philadelphia Academy of Natural Sciences, established in 1812, and in 1817 he was asked to edit the academy's newly founded *Journal*. In the same year, he published a prospectus for his projected *American Entomology*, which was to cover all known American insects, many of them illustrated with colored plates. The project was resumed in 1824, but never brought to completion.

Say's acquaintances included botanist Thomas Nuttall, ornithologist Alexander Wilson, and zoologist George Ord. In 1817 and 1818, Say and Titian Peale, son of artist Charles Willson Peale, founder of Peale's Museum, with George Ord and the wealthy William Maclure, undertook a collecting trip to Florida that terminated suddenly when there was a threat of Indian attack. Say was a modest person; in George Ord's words: "His disposition was so truly amiable, his manners were so bland and conciliating, that no one, after having once formed his acquaintance, could cease to esteem him." In 1819 Say was thirty-two.

Long instructed that "Mr. Say will examine and describe any objects in zoology, and its several branches, that may come under our observation. A classification of all land and water animals, insects, etc., and a particular description of the unusual remains found in a concrete [fossilized] state will be required of him." In fact, study of the customs of the Indians, originally assigned to Biddle, fell to Say, who earned a considerable reputation as an ethnologist. But his first love was insects, and he is now remembered best as the first American systematic entomologist.

Say's good friend Titian Ramsay Peale, who joined the expedition as assistant naturalist, was only nineteen at the time, but he was experienced in taxidermy and came from a family of artists. His father, Charles Willson Peale, not only was an artist of note, but had founded the first natural history museum in the country. He married three times and had sixteen sons and daughters, not all of whom survived beyond childhood. Several of his sons were named for artists: Raphaelle, Rembrandt, Rubens, and Titian. Peale's Museum contained, by 1802, more than 1,800 specimens of birds, 250 mammals, hundreds of fishes and reptiles, and thousands of insects. Zebulon Pike supplied two young grizzly bears—one of the few natural history legacies of his expedition. Soon the bears became too difficult to handle and had to be killed and stuffed, but not before Titian had painted them in watercolors, one of his first animal paintings. Charles Willson Peale had visited Secretary of War Calhoun in 1818, successfully urging him to include Titian as a member of the expedition and to use his museum as a major depository for specimens collected in the West.

Titian Peale had been a rebellious youth, at least in the opinion of his overbearing father. In 1816 Charles Willson Peale had written to his son Rubens: "I am really fearful that he will become dissipated and a disgrace to the family." His efforts to train his son to a trade were in vain; Titian was a born naturalist, already collecting and painting insects. In November 1817, when he was only eighteen, he was elected the youngest member of the Academy of Natural Sciences of Philadelphia, partly on the basis of the illustrations he was doing for Say. A month later, he was off to Florida with Say, where he showed himself to be an energetic collector of birds, mammals, insects, and Indian artifacts. On the Long Expedition, Peale was to prove himself an excellent wildlife artist and a good shot with a rifle as well as a dependable preparator of specimens.

Augustus Jessup was a prosperous Philadelphia businessman with an interest in natural science, but with no solid qualifications as a geologist. He was expected to study "earth, minerals, and fossils, distinguishing the primitive, transition, secondary, and alluvial formations and deposits." Jessup left the expedition after six months. His duties as geologist were assumed by Edwin James during the trip to the Rockies in 1820.

"In this science [geology], as also in botany and zoology," Long added, "facts will be required without regard to the theories or hypotheses that have been advanced on numerous occasions by men of science." This seems a curious directive. Was he asking his "scientific gentlemen" to avoid reflection on the significance of their findings? Perhaps he looked with disfavor on Linnaeus's classification of plants according to their sexual organs, or on Lamarck's recent heresies concerning the evolution of higher forms of life from lower.

More probably, he was simply reflecting the philosophy of Francis Bacon, still influential nearly two centuries after his death. In his *Novum Organum* (New Instrument), Bacon extolled the importance of inductive—building theories from facts—as opposed to deductive reasoning. Thomas Jefferson was a strong Baconian, protesting against "visionary theories," and doubtless Long's training and his contacts with members of the American Philosophical Society had colored his thinking.

Long also wanted an experienced artist to "furnish sketches of landscapes, whenever we meet with any distinguished for their beauty and grandeur. He will also paint miniature likenesses, or portraits, if required, of distinguished Indians, and exhibit groups of savages engaged in celebrating their festivals, or sitting in council, and in general illustrate any subject, that may be deemed appropriate to his art." In March 1819, Long requisitioned from a Philadelphia emporium a variety of artist's supplies, including drawing paper, pencils, brushes, and diverse watercolors.

Samuel Seymour had emigrated from England and become associated with Thomas Sully and other Philadelphia artists. He was a member of the Columbian Society of Artists, and had entered several of the society's exhibitions. In 1819 he was about thirty-five years of age. Evidently his work appealed to Long. To portray for the first time the vast western landscapes would have been a challenge to any artist. Seymour's painting *Distant View of the Rocky Mountains*, which formed the frontispiece of the expedition's 1823 report, provided the public with its first glimpse of that almost mythical range. His (and Peale's) renditions of western Native Americans also antedated by more than a decade the better known paintings by George Catlin and Karl Bodmer.

A few months before their departure, Charles Willson Peale painted portraits of the explorers, remarking that "if they lost their scalps, their friends would be glad to have their portraits."

$\mathscr{T}hree$
DOWN THE OHIO

STEPHEN LONG SUPERVISED THE CONSTRUCTION OF the steamboat *Western Engineer* in Pittsburgh during 1818 and early 1819. It was originally planned as a smaller craft, drawing only fourteen inches of water, but as finally completed, it drew more than two feet of water and was unlike any vessel that had plied the Ohio before. The ship left the arsenal on the Allegheny River with much fanfare on the afternoon of May 3, a Monday, but did not begin to descend the Ohio until Wednesday. I will let Titian Peale describe the departure and the ship itself. Allowance must be made here as elsewhere for the fact that in the early 1800s people were not always fastidious in their spelling and punctuation.

MAY 3, 1819. TP: Left the garrison 2 miles from Pittsburg on the Alleghany River at 4 o'clock in the afternoon after firing a salute of 22 guns which were answered with as many from the arsenal. As we steered for Pittsburgh our appearance attracted great numbers of spectators to the banks of the River. We fired a few guns and were cheered in return from the shore. Our boat appears to answer very well, but being quite new, the machinery is rather stiff. . . . She draws about two feet and a half water, the wheels placed in the stern in order to avoid

Titian Peale, ink-and-wash sketch of the steamboat *Western Engineer*. (American Philosophical Society)

trees, snags and sawyers, etc. On the quarter deck there is a bullet proof house for the steersmen. On the right hand wheel is *James Monroe* in capitals, and on the left, *J. C. Calhoun,* they being the two propelling powers of the expedition. She has a mast to ship and unship at pleasure, which carries a square and topsail, on the bow is the figure of a large serpent, through the gapping mouth of which the waste steam issues. It will give, no doubt, to the Indians an idea that the boat is pulled along by this monster. Our arms consist of one brass four pounder mounted on the bow, four brass 2⅞ inch howitzers, two on swivels, and two on field carriages, two wolf pieces carrying four ounce balls; twelve muskets, six rifles, and several fowling pieces, besides an air gun, twelve sabers, pistols, and a quantity of private arms of various sorts and a great sufficiency of ammunition of all kinds for our purpose. This evening, we sent up a few rockets.

MAY 4. TP: We tried the stream and took a few turns on the river, and were visited by . . . gentlemen who advised many alterations. . . . We were visited today by a commissioner of the Bible Society who left us two bibles and one or two other books for the good of our souls. . . .

Days spent waiting for the departure of the expedition were not wholly wasted. Using a hook and line, the naturalists dredged from the Allegheny River a curious salamander, ten inches long, with external gills: a mud puppy. Edwin James's *Account* includes, as a footnote, a description of this animal by Thomas Say, who named it *Triton lateralis* in the belief that it differed significantly from other known salamanders. In fact, Constantine Rafinesque had discovered and named it a year earlier, so it is Rafinesque's name, *Necturus maculosus*, that is now used. Rafinesque, the most colorful and eccentric of the frontier naturalists, had been describing animals and plants of the Ohio Valley at a prodigious rate, and more than once "scooped" Thomas Say.

While waiting, Say also took occasion to write to friends concerning his accommodations on the *Western Engineer*. His cabin, he reported, was small but adequate, with space for a table and for a shelf of natural history books. His bookshelf housed the works of Linnaeus and of the Danish entomologist J. C. Fabricius, along with twelve volumes of William Nicholson's *British Encyclopedia* and other publications.

Say, Peale, and William Baldwin had traveled before, but never as part of a military operation and never with the fanfare of this departure from Pittsburgh. They can have had little perception of what lay ahead—indeed, the expedition's second year was at that time wholly unplanned. Peale's journal remains dry and factual, yet the young man—only nineteen—must have been at least slightly apprehensive, as indeed must all the men have been.

MAY 5. TP: Having completed all alterations and taken all stores aboard at ½ past 4 in the afternoon we bid adieu to Pittsburgh and descended rapidly down the Ohio. At about fourteen miles below the town we saw a steam boat grounded. We received and returned her salute as we passed by. In the evening we heard the first cry of the Whipoorwill (*Caprimulgus vociferus*). Vegetation is progressing very rapidly. Most of the forests are already clothed. In coming down the river saw a Cormorant . . . and two Turkey Vultures. We saw some bird that I took to be the Tell-tail Sandpiper [an early name for the greater yellowlegs]. Our boat seems to attract universal attention, the people stopping all along the shore to gaze at us as we pass by.

James's description of the departure in the official report of the expedition follows that of Peale almost word for word, indicating that here as elsewhere he had access to Peale's journal. Before dawn, on May 6, the *Western Engineer* reached Steubenville, Ohio. While the crew loaded wood to fuel the engines, Peale toured the town, which he said "contains many houses that would not discredit Philadelphia." There were fox squirrels in the trees. A few hours later, they passed Wheeling.

MAY 6. TP: Soon after [passing Wheeling] we experienced a very violent storm from the S.W. accompanied with thunder and the heaviest I have ever experienced this season. The country we are passing through is grand and beautiful in the extreme. Vegetation appears farther advanced than higher up the river, saw the first Humming Bird (*Trochilus colubris*) that I have seen this season blown over the boat in the storm [this was a ruby-throated hummingbird, now *Archilochus colubris*]. Saw, also, the small green heron (*Ardea virescens*) for the first time on this river [this was presumably a green-backed heron, now *Butorides striatus*].

Peale tells us that they "passed Charleston," firing a gun as a salute. However, James states that they made "an excursion on shore, near the little village of Charleston, in Virginia." Reuben Gold Thwaites, in his 1905 reprinting of James's *Account*, identifies the village as Charleston, the present capital of West Virginia, which is on the Kanawha River many miles from the Ohio. Surely this cannot be correct (James states that they passed the *mouth* of the Kanawha on May 7). Wherever this Charleston may have been—presumably on the Ohio only a short distance below Steubenville (they left Steubenville at ten in the morning and reached Charleston around noon)—James has something to say concerning the "excursion on shore," extracted from Baldwin's notes.

MAY 6. EJ: [W]e met with many plants common to the eastern side of the Alleghanies; beside the delicate sison bulbosum, whose fruit was now nearly ripened [this is an early-flowering member of the carrot family, called harbinger-of-spring, now known to science as *Erigenia bulbosa*]. In shady situations we found the rocks, and even the trunks of trees to some little distance from the ground, closely covered with the sedum ternatum [a stonecrop], with white flowers fully unfolded. The cercis canadensis [redbud], and the cornus florida [flowering dogwood], were now expanding their flowers, and in some places occurred so frequently, as to impart their lively colouring to the landscape. In their walks on shore, the gentlemen of the party collected great numbers of the early-flowering herbaceous plants, common to various parts of the United States.

James here includes a long footnote listing many of the plants observed, along with their dates of blooming. The naturalists were still in relatively well explored country, and there seems little point in repeating this list. Redbud and flowering dogwood, for example, had been supplied with scientific names by Swedish naturalist Carl Linnaeus, on the basis of samples sent to him from America. *Sison bulbosum* and *Sedum ternatum* had been described by André Michaux. Michaux, a well-trained and widely traveled botanist, was sent to the United States in 1785 to find plants useful for French gardens. He had been as far west as Illinois and was the first naturalist to collect plants on the prairies. Michaux returned to France in 1796 after discovering and describing innumerable plants from the eastern half of the country in his *Flora Boreali-Americana*. The naturalists of the Long Expedition would find few botanical novelties until they reached the Great Plains.

> EJ: The scenery of the banks of the Ohio, for two or three hundred miles below Pittsburgh, is eminently beautiful.... Broad and gentle swells of two or three hundred feet, covered with the verdure of almost unbroken forest, embosom a calm and majestic river; from whose unruffled surface, the broad outline of the hills is reflected.... These forests are now disappearing before the industry of man; and the rapid increase of population and wealth, which a few years have produced, speak loudly in favour of the healthfulness of the climate, and of the internal resources of the country....

On May 7, the *Western Engineer* stopped at Marietta, Ohio, for wood, and the next day a wood stop was made near Gallipolis, where Peale found "the finest beech woods I ever saw." Early on May 9, the expedition stopped for wood at Maysville. Each of these daily wood stops took several hours, sometimes as many as eight, affording the naturalists an opportunity to explore the shore but delaying the expedition to an extent that can hardly be imagined today. At Cincinnati, which they reached late on May 9, there was to be a delay of more than a week.

> MAY 9. TP: Cincinnati, like some other towns in the western country, had risen like a mushroom from the wilderness. . . . The present population is said to exceed 25,000. Immigrants are every day arriving from all parts of the world. The inhabitants have already founded a college and subscribed eight or ten thousand dollars for a museum. They have a few articles collected for that purpose, mostly fossils and animal remains. Wishing to make some alterations in the machinery of the boat, and Dr. Baldwin being very sick, it was determined to stay here several days. The doctor has accordingly been moved on shore to the house of Mr. Glenn and Dr. Drake summoned to attend him.

The "Dr. Drake" to whom Peale refers was Daniel Drake, one of the leading citizens of Cincinnati. He had been brought up on the frontier and had a somewhat makeshift education, but when he was only fifteen he was apprenticed to the eccentric Dr. William Goford of Cincinnati, whose medical practice he later took over. By 1819, Drake had founded a hospital, a library, a college, and the museum to which Peale refers. He was a prolific writer and ever ready to boost Cincinnati as the "Philadelphia of the West."

Drake's Western Museum was small, but nevertheless one of the major attractions of the city. As his preparator, Drake had hired a thirty-three-year-old painter and bird enthusiast of French extraction. Peale does not mention him—oddly, since he himself came from a family of artists and was no mean artist himself. But John James Audubon did remember how "Messrs. T. Peale, Thomas Say and others stared at my drawings." Were they not impressed?

> EJ: Vegetation is here luxuriant; and many plants unknown eastward of the Alleghany mountains, were constantly

presenting themselves to our notice. Two species of aesculus are common. One of these has a nut as large as that of the ... common horse-chestnut of the gardens.

These nuts are round, and after a little exposure become black, except that part which originally formed the point of attachment to the receptacle, which is an oblong spot three-fourths of an inch in diameter; the whole bearing some resemblance to the eyeball of a deer, or other animal. Hence the name *buck-eye*, which is applied to the tree [Ohio buckeye, *Aesculus glabra*]. The several species of aesculus are confined principally to the western states and territories. In allusion to this circumstance, the indigenous backwoodsman is sometimes called buck-eye, in distinction from the numerous emigrants who are introducing themselves from the eastern states. The opprobrious name of Yankee is applied to these. . . .

MAY 18. EJ: [T]he weather becoming clear and pleasant, Dr. Baldwin thought himself sufficiently recovered to proceed on the voyage; accordingly, having assisted him on board the boat, we left Cincinnati at ten o'clock. . . .

Below Cincinnati the scenery of the Ohio becomes more monotonous than above. . . . This is, however, somewhat compensated by the magnificence of the forests themselves. Here the majestic platanus [sycamore] attains its greatest dimensions, and the snowy whiteness of its branches is advantageously contrasted with the deep verdure of the cotton-wood, and other trees which occur on the low grounds. . . .

The fruit of the sycamore is the favorite food of the paroquet, and large flocks of these gaily-plumed birds constantly enliven the gloomy forests of the Ohio.

From this date, there were several sightings of Carolina parakeets (or paroquets). These, the only members of the parrot clan occurring in the United States aside from the deep Southwest, were about a foot long (more than half of that tail) and were bright green with a yellow and orange head. They moved about in flocks and were an easy mark for plumage hunters and for fruit growers who resented their forays into their orchards. When John Kirk Townsend passed through Missouri in 1834, parakeets were still abundant:

> They seem entirely unsuspicious of danger, and after being fired at, only huddle closer together, as if to obtain protection from each other, and as their companions are falling about them, they curve down their necks, and look at them fluttering upon the ground, as though perfectly at a loss to account for so unusual an occurrence. It is a most inglorious sort of shooting; down right, cold-blooded murder.

The last living parakeet was seen in 1904, but as late as 1934 Roger Tory Peterson wrote that "naturalists still hope that a stray individual or flock might turn up." The expedition members, of course, could not dream that these spectacular birds would some day no longer be part of the American scene.

After leaving Cincinnati, Long attempted to make up for lost time by traveling all night, and Louisville, 120 miles downstream, was reached the next day. During the night, the *Western Engineer* passed the ships of the military contingent, which had left Pittsburgh before it. At Louisville, the expedition remained for four days while further repairs were made on the engines. When the boat left, a pilot was taken on board to conduct it through the "falls of the Ohio," a drop of twenty-two feet in less than two miles. "The water boils and splashes about in a most violent manner," wrote Peale, "and in one place resembles exactly the surf of the sea." The ship nevertheless passed through the rapids without problems.

At the foot of the falls was the town of Shippingport, Kentucky (not to be confused with Shippingport, Pennsylvania, which is not far from Pittsburgh). There, larger boats ascending the Ohio often waited for high water before challenging the rapids, or portaged their cargo to Louisville. "A few days ago," wrote Peale, "there was no less than twenty steamboats unloading here, most of them in the New Orleans trade." It was in Shippingport that Audubon had lived for a time with his wife's relatives after his business had failed and before he moved to Cincinnati to serve as Drake's preparator.

James listed some of the plants common in the Louisville area, doubtless extracted from Baldwin's journals (Baldwin rarely left his sickbed, relying on others to bring him specimens). Included were species of Saint-John's-wort, nightshade, milkwort, and several grasses. The pastures, he noted, were "much overrun with luxuriant weeds," particularly jimsonweed (*Datura strammonium*) and mayweed (*Anthemis Cotula*).

After taking on wood, the ship proceeded downstream from the rapids on May 23. Near the mouth of the Wabash River, the engine failed, and the ship was allowed to drift until it arrived at a place where repairs could be made. This allowed the naturalists time to go on shore. Peale saw wood ducks, a gull, and "a number of little Marsh Terns (*Sterna minuta*)" (probably least terns, *S. antillarum*). Least terns, which are hardly larger than swallows, are now listed as endangered throughout the central states, as the sandy beaches on which they nest have largely been flooded or trampled.

On the following day, Peale went hunting and killed a turkey and saw a deer, many gray squirrels, and several pileated woodpeckers. On a lake he found a turtle depositing its eggs in the sand. He called it the "Lake Erie Tortoise (*Testudo geographica*)." Today we would call it the map turtle (*Graptemys geographica*), named for the intricate patterning of lines on the shell.

MAY 28. TP: Killed several specimens of the Little Tern . . . and five Semipalmated Sandpipers. . . . The Tern appears to

DOWN THE OHIO

Titian Peale, watercolor of a least tern, May 28, 1819. (American Philosophical Society)

be attracted here by great numbers of a species of Phryganea [a caddisfly] with which I found the stomach of one I opened filled. The Semipalmated Sandpipers were in pretty large flocks and did not appear stationary [resident] here.... Proceeded at 2. At 4 went aground on a sandbar.... By dint of anchor, setting poles, steam, and all of the men in the water prying her, we got off just at dark, and ran down hill until we were opposite a cave in the rock where we laid up for the night. Next morning we visited the cave.

There follows a description of the cave, which extended for 160 feet into the limestone cliffs. On the cliff top, a pair of raptors was sighted, believed to be red-shouldered hawks. Shortly after leaving, the ship once again "ran aground on a sandbar and did not get off until after eleven o'clock.... Lay to soon after in a thunderstorm. There was the most vivid lightning I ever saw, being awfully grand," wrote Peale. James's *Account* here includes a long footnote describing the mineralogy of the area, taken from Jessup's report. Much of it concerns the availability of salt at sites along the Ohio and else-where. Salt was essential to settlers for the preservation of meats, fish, and other foodstuffs.

MAY 29. EJ: On [this day] we passed the mouths of the Cumberland and Tennessee, the two largest rivers, tributary to the Ohio. At the mouth of the Cumberland is a little village called Smithland. . . . [We'll hear of Smithland again; James was to spend the winter of 1820/1821 there, ill and without the money to return to Philadelphia.] About half way between the mouth of the Cumberland and Tennessee . . . are several large catalpa trees. They do not, however, appear to be native. . . . [The name] may be a corruption from Catawba, the name of the tribe by whom, according to the suggestion of Mr. Nuttall, the tree may have been introduced.

The catalpa trees that the naturalists observed were very probably native, as they were now within the range of the western catalpa (*Catalpa speciosa*). Most authorities believe that the name "catalpa" is based on Creek Indian words meaning "head with wings," with reference to the shape of the showy flowers. The flowers are followed by long, pendant seedpods, hence the name "cigar tree" sometimes applied to these trees. Note that James continually reports what "we" did, even though he did not join the expedition until the following spring. He was, of course, speaking for the present members of the expedition, to whose notes and diaries he had access.

MAY 30. EJ: In the afternoon of the 30th we arrived at the mouth of the Ohio. This beautiful river has a course of one thousand and thirty-three miles, through a country surpassed in fertility of soil by none in the United States. Except in high floods, its water is transparent, its current gentle. . . . The lands about the junction of these two great rivers are low, consisting of recent alluvion, and covered with dense forests. At the time of our journey, the spring floods having subsided in the Ohio, this quiet and gentle river seemed to be at once swallowed up, and lost in the rapid and turbulent current of the Mississippi.

MAY 30. TP: Saw on the Ohio just above its mouth a Fish Hawk (*Falco Haliaetus*) [now *Pandion haliaetus*, osprey] and a bird at a distance that I supposed to be the White Pelican. Came to about two miles above the mouth of the Ohio on the eastern bank of the Miss. Went ashore but found it almost impossible to hunt, the mosquitos being so numerous and immense quantities of nettles making it painful to walk in the woods.

It had taken the expedition nearly a full month to complete the descent of the Ohio. Aside from frequent stops for wood, there were several occasions when the *Western Engineer* became stuck on sandbars or had to stop because of engine problems. Baldwin's illness had kept the explorers in Cincinnati for ten days. Baldwin, Say, Peale, and Jessup must have felt frustrated that what was no more than a preamble to their expedition had taken so long. Now they were at last on the Mississippi, but would have to travel another ten days to reach St. Louis and the mouth of the Missouri River.

MAY 31. EJ: Finding it necessary to renew the packing of the piston in the steam engine, which operation would require some time, most of the gentlemen of the party were dispersed on shore in pursuit of their respective objects, or engaged in hunting. . . . We were gratified to observe many interesting plants, and among them several of the beautiful family of the orchidae, particularly the orchis spectabile [showy orchid], so common in the mountainous parts of New England.

Peale and Jessup found camps of Pawnee Indians, at one of which they purchased a deer. There were many tracks of wild turkeys; yellow-breated chats and parakeets were common. Bank swallows and kingfishers nested in holes along the riverbank. In the evening, the explorers were serenaded by chuck-will's-widows. On June 5, the sail was hoisted and the ship moved upstream rapidly, passing Kaskaskia, Illinois, a town that had been founded fully a century earlier by Jesuits, who named it after the local Indian tribe. André Michaux had used the village as a center for botanizing in 1795. Just beyond, the *Western Engineer* passed Ste. Genevieve, on the Missouri side, and then suddenly hit a snag that caused a major leak.

JUNE 5. TP: Set all hands to pumping and came to at dusk, on the western side, the hands at the pump all night.

JUNE 6. TP: Discovered the leak to be in the stern just below the water mark; having lightened her we erected a pair of shears and raised her stern sufficiently to get at the leak and caulk it. Opposite the place where we lay on a sandbar was a large flock of White Pelicans which remained in the same place all the morning. Started at 11 o'clock and passed some of the most sublime bluffs of limestone rocks that I ever beheld. . . .

On June 7, the boat took on wood but soon could not advance against the current because of an accumulation of mud in the boilers. To remove the mud, it was necessary to cool the boilers and crawl inside to clean them; then the boilers had to be refilled and the furnace restarted—a matter of several hours in all. During the delay, the naturalists explored the shore and collected a rat that Peale described as "very ferocious." Examination of the stomach showed it to be filled with green bark and plant shoots. The nest was composed of great quantities of brush and detritus. This was a pack rat, more precisely an eastern wood rat, now called *Neotoma floridana.* Among the plants collected for Baldwin's study were samples of two trees—

smooth sumac and common persimmon—and wildflowers, including species of phlox, aster, bedstraw, and others.

On June 9, the *Western Engineer* arrived in St. Louis, where Peale reported that it was received "with a salute from a 6 pounder on the bank and from several steam boats along the town." The next day, the officers and scientists were treated to a banquet to which the officers of the military contingent and "all the captains of the steamboats in port" were invited. The unusual design of Long's ship attracted much attention and elicited these remarks from the St. Louis *Enquirer*:

The bow of the vessel exhibits the form of a huge serpent, black and scaly, rising out of the water from under the boat, his head as high as the deck, darted forward, his mouth open, vomiting smoke, and apparently carrying the boat on his back. From under the boat, at its stern issues a stream of foaming water, dashing violently along. . . . Neither wind nor human hands are seen to help her; and to the eye of ignorance the illusion is complete, that a monster of the deep carries her on his back smoking with fatigue, and lashing the waves with violent exertion.

St. Louis had been founded by the French in 1764 and named in honor of Louis XV, but the city had no sooner been platted than Louisiana was ceded to Spain, only to be returned to France in 1801 and finally sold to the United States in 1803. In 1819, one was as likely to hear French, Spanish, or Indian dialects on the street as English. William Clark was superintendent of Indian affairs, while Manuel Lisa (of Spanish descent) and Auguste and Pierre Chouteau (of French descent) dominated the fur trade. The city had become the gateway to the West, where fur traders, military personnel, and settlers completed their preparations for journeys into lands occupied primarily by Native Americans. Furs were shipped east and often on to Europe via the Ohio River or New Orleans. The city had a population of about 4,000.

Say and Peale spent several days studying Indian mounds in the St. Louis area. James's report includes detailed measurements of twenty-seven of these mounds, taken from notes made by Say, who found bones, pottery, and other objects in some of the graves he opened. James was once again moved to florid language.

> EJ: A survey of these productions of human industry, these monuments without inscription, commemorating the existence of a people once numerous and powerful, but no longer known or remembered, never fails, though often repeated, to produce an impression of sadness. As we stand upon these mouldering piles, many of them now nearly obliterated, we cannot but compare their aspect of decay with the freshness of the wide field of nature, which we see reviving around us; their insignificance, with the majestic and imperishable features of the landscape. We feel the insignificance and the want of permanence in every thing human; we are reminded of the pyramids of Egypt, and may with equal propriety be applied to all the works of men, "these monuments must perish, but the grass that grows between their disjointed fragments shall be renewed from year to year."

"The grassy plains to the west of St. Louis are ornamented with many beautifully flowering herbaceous plants," says James at another point. Among those studied by Baldwin were yellow lady's slipper, southern red lily, Dutchman's pipe, scarlet Indian paintbrush, spiderwort, rock rose, and clematis.

> EJ: The borders of this plain begin to be overrun with a humble growth of black jack [oak] and the witch hazel, it abounds in rivulets, and some excellent springs of water, near

one of which was found a new and beautiful species of viburnum. On the western borders of this prairie are some fine farms.
It is here that Mr. John Bradbury, so long and advantageously
known as a botanist, and by his travels into the interior of
America, is preparing to erect his habitation. This amiable gentleman lost no opportunity during our stay at St. Louis to make
our residence there agreeable to us. Near the site selected for
his house is a mineral spring, whose waters are strongly im-
pregnated with sulphuretted hydrogen gas. Cattle and horses,
which range here throughout the season, prefer the waters of
this spring to those of the creek in whose bed it rises, and may
be seen daily coming in great numbers, from distant parts of
the prairie, to drink of it.

John Bradbury's *Travels in the Interior of America*, describing his experiences on the Missouri in 1810 and 1811, had been published just
two years earlier. Bradbury had been commissioned by the Botanical
Society of Liverpool to collect plants suitable for cultivation in England, but he soon became engrossed in the unstudied flora of the
West. He and fellow botanist Thomas Nuttall traveled with the Astorians as far as Fort Mandan (North Dakota). Bradbury returned to
England hoping to describe the plants he had sent back, only to find
that his "design was frustrated, by my collection having been submitted to the inspection of a person by the name of Pursh," who
included the descriptions in his *Flora*. Bradbury returned to settle in
Missouri in 1818.

Despite the opportunity to visit with Bradbury, the naturalists
must have grown impatient with Long's twelve-day delay in St.
Louis. He apparently waited until Colonel Henry Atkinson's troop-
laden steamboats were ready to proceed up the Missouri, and there
were problems obtaining sufficient supplies to sustain the troops
through the winter. Neither Atkinson nor Long had appreciated the
many problems in navigating the rivers in steamboats—and there

were more problems to come. Atkinson decided to overwinter his troops at Council Bluff, some 600 miles upstream, rather than trying to go on to the mouth of the Yellowstone as originally planned. Long changed his plans to follow suit. Finally, on June 21, the *Western Engineer* left St. Louis and on the following day entered the Missouri.

Four

UP THE MISSOURI

JUNE 22. TP: The meeting of the waters of the Mississippi and Missouri have a very singular appearance. The Mississippi is clear and of a transparent green, the Missouri thick with yellow mud and being heavier than the Mississippi, the water takes the bottom of the river until it strikes the opposite shore. It is there thrown to the surface and presents large spots of muddy water intermingled with the clear. The Minute Terns were very numerous near the confluence of the two rivers. This morning heard the cries of a flock of Parrakeets. Arrived opposite Bellefontaine [four miles up the Missouri] at 2 o'clock and ran on a sand bar which kept [us] until 4 before we landed. The 6th regiment are encamped on the banks of the river opposite the garrison of Bellefontaine. Col. Atkinson is contriving his boats to go by wheels turned by the soldiers. Each boat is to have two pairs of wheels and 8 men at each pair.

Only once before had steamboats plied the turbulent Missouri, and then for only a short distance. Even with the use of manpower at the side wheels, Colonel Atkinson's troop ships were not succesful in ascending the Missouri very far. Two, the *Calhoun* and the

Exchange, turned around almost immediately; another, the *Jefferson*, a few days later. The *Johnson's* machinery gave out ten miles below the mouth of the Kansas River, and only the *Expedition* reached Camp Martin, on Cow Island, near the mouth of the Kansas. William D. Hubbell, a clerk on the *Johnson*, recollected that on many nights those on the boat "could look back and see the place we started from in the morning. . . . [T]he whole matter of the steamboats was as complete a failure as could have been possible, the boats being totally unfit for the trip." The troops who finally reached their winter quarters several miles above Council Bluff did so primarily on keelboats. So much for the hope of reaching the Yellowstone, many hundreds of miles beyond!

The *Western Engineer*, with its narrow beam and rear paddle wheel, was better adapted to the Missouri. Nevertheless the "thick yellow mud" of the river accumulated rapidly in the boilers. Rather than cleaning them out manually, Stephen Long and his crew devised a method of blowing out the mud through a tube by steam pressure. Progress was, however, extremely slow.

> JUNE 24. TP: Left Bellefontaine at 10 o'clock. Proceeded up very slowly at first owing to the rapidity of the current. In one place the boat could scarcely make one mile per hour. Calculating that she would go 6 in still water, it would make the current run at the rate of 5 miles per hour. . . . Grounded twice in the afternoon but cleared ourselves without much difficulty. The Missouri during today's run was much obstructed by sandbars and islands. . . . The Yellow Breasted Chat (*pipra polyglotta*) [now *Icteria virens*] is extremely common, both in the prairies [and] where there are bushes and woods.

Yellow-breasted chats were doubtless more common than they are now, as many of the thickets in which these idiosyncratic warblers breed have been removed for agriculture. In late June, the chats were probably still in song—if the curious hoots and whistles of these birds can be called a song. The naturalists were surprised to see "black-headed terns" so far from the ocean. These may well have been Forster's terns, which breed in marshes in the interior of the West. Basswood trees (*Tilia americana*) grew large and were in full flower. There were wild roses and a species of milkvetch that Baldwin thought might belong to a new genus. James noted that sandbars formed at the bends in the river were quickly colonized by small willows and cottonwoods.

Now that they were in country where the insect fauna was poorly known, Say was collecting insects that he later named and described. Several times he noted migrations of a small, brightly colored leaf beetle, which he named *Altica 5-vittata* (now *Disonycha 5-vittata*). A planthopper (*Delphax tricarinata*, now *Stobaera tricarinata*), "came on board . . . in considerable numbers." A small shore bug was "not uncommon on the shore of the Missouri River, skipping nimbly about." Say named it *Acanthia interstitialis*, unaware that the species also occurred in Europe and had been named in 1794 by Danish entomologist J. C. Fabricius. It is now called *Saldula pallipes*. As is often the case, all three species are now placed in genera of more restricted scope than those in which Say placed them.

At St. Charles, Major Benjamin O'Fallon, the Indian agent, and his deputy and interpreter, John Dougherty, joined the expedition. From now on there would be many contacts with Indians, and it was hoped that these two experienced men would help in dealing more effectively with them. Meanwhile, tension was growing between Biddle and Long. The two refused to speak to each other, and Biddle challenged Long to a duel, which fortunately never came to pass.

Long now decided to send a group headed by Say overland, to rejoin the ship farther upstream. They were to be away for a week while the *Western Engineer* labored up the river. One evening Baldwin felt well enough to walk on shore, "but returned much fatigued by his exertions." Baldwin listed some of the "vegetable productions at this place." Dominant trees included cottonwood, sycamore, hickory, red oak, American hornbeam, sassafras, and juniper. Climbing bittersweet was in fruit, and orange coneflowers (*Rudbeckia fulgida*) were in bloom.

JUNE 27. EJ: The shore here was lined with the common elder, (sambucus canadensis) in full bloom, and the cleared fields were yellow with the flowers of the common mullein. This plant, supposed to have been originally introduced from Europe, follows closely on the footsteps of the whites. [Great mullein (*Verbascum thapsus*) had indeed come from Europe, where it had a variety of uses in folk medicine, including the alleviation of asthma, earache, and even bed-wetting.] The liatris pycnostachia, here called "pine of the prairies" [prairie snakeroot or blazing star], which was now in full bloom, has a roundish tuberous root, of a warm somewhat balsamic taste, and is used by the Indians and others for the cure of gonorrhoea.

The Indian interpreter, Mr. Dougherty, also showed us some branches of a shrub, which he said was much used among the natives in the cure of lues venerea [syphilis]. They make a decoction of the root, which they continue to drink for some time. It is called "blue wood" by the French, and is the symphoria recemosa [a species of snowberry, *Symphoricarpos racemosus*] . . .

Syphilis was evidently prevalent among Native Americans. The first major outbreak in Europe occurred soon after Columbus's return, and it is often maintained that some of his sailors brought back the spirochetes from the West Indies. The disease may have been endemic among tribes of the American West, or they may have become infected through contacts with early trappers and traders of European descent. Gonorrhea likely had its origins in the Old World, as it was known in ancient times. It was probably delivered to Native Americans along with smallpox and a variety of other diseases.

EJ: Without meeting any remarkable occurrences, we moved on from day to day, encountering numerous obstacles in the navigation of the river, and being occasionally delayed by the failure of some part of the steam-engine, till on the 2d of July, we arrived at Loutre [Otter] Island, where we found Mr. Say and his companions.

The group that had traveled overland seems to have encountered many difficulties without accomplishing very much. Say, Peale, Jessup, and Seymour left the ship at St. Charles on the morning of June 26, buying a horse and loading it with their tent, blankets, canteens, and "a few biscuits." A man named Kenna came along to lead the horse. The first day they traveled eighteen miles over a shadeless prairie, with water in short supply. Their horse threw its pack and ran off, but with some effort was recovered. In the evening, they camped not far from a house, but the owner would not give or sell them anything to eat and gave them water "with a bad grace." They made supper of a hawk that Jessup had shot.

Fortunately the next day's travel brought them to a house where they were able to obtain water, milk, and cornbread, and in the evening they arrived at "Kenedy's fort," where they camped and "made a supper on some squirrels, and larks, to which we added some corn bread and butter milk that we had purchased from Kenedy." Kenedy also sold them a ham, which they consumed the next day. The prairie was filled with "partridges and larks" and "some few of Bartram's Sandpiper (*Tringa Bartramia*)" (now *Bartramia longicauda*, upland sandpiper). These relatively long-necked sandpipers nest on the ground in tall-grass prairies, migrating in the fall to southern South America. Was Say aware at the time that they were seeing a bird that had been named for his great-grandfather?

JUNE 29. TP: Kenna returned at sunrise without the horse or having seen anything of her. On the prairies there is a species of green headed fly [horse fly, *Tabanus*] which torments horses and cattle so much that crossing the prairies in the day is next of impossible. This, I believe, was the cause of our horse running away. It is said that these flies will sometimes kill a horse, therefore in crossing prairies travelers mostly turn night into day and day into night.

No alternative was left to us but to divide the pack and turn pack horses ourselves, then to make to the nearest point on the river. . . . Arrived at Loutre Island just at dark. Here we found water in plenty, and it seemed as though it was impossible for us to satisfy our thirst. . . . Distance 21 miles, but rough roads, a hot sun, and heavy loads, made it the longest 21 miles ever I traveled.

After breakfasting, struck our tent, and marched up the river 2 miles. Came to another house on the island and encamped on the banks of the river. In the march I killed 4 turkeys, two of them at one shot. They are more numerous here than ever I saw them. Procured a kettle at the house, cooked, and feasted on our turkey.

JULY 1. TP: Went out to hunt, killed a turkey, a rabbit, and some squirrels which are extremely numerous. A person walking through the woods is scarcely ever out of sight of them. Parakeets abound; deer are numerous. The land is very rich, and covered with remarkably fine timber. . . .

JULY 2. TP: Went out to hunt, killed 2 rabbits, a polecat, and wounded a fine buck. Mr. Jessup killed a turkey. In the evening we were gratified by the approach of our boat. She passed us and came to a mile above where we all went on board. . . .

Loutre Island was large enough to have several farms, and Long was able to obtain a supply of fresh vegetables, eggs, and poultry. After describing the farm buildings and corn mills, the naturalists commented on some of the lightning beetles they had observed: "A large

species of lampyris is common on the lower part of the Missouri. . . .
It emits from three to seven or eight flashes, in rapid succession, then
ceases; but shortly after renews its brilliancy." At Loutre Island this
species was absent, but had been replaced by "great numbers of the
lampyris pyralis, whose coruscations are inferior in quantity of light,
and appear singly." In recent years there has been much research on
the flashing of the males of these beetles, demonstrating that each
species has its own pattern of flashes to which females of that species
respond. It is interesting that members of the expedition were alert
to the differences between these two species.

> EJ: The black walnut attains, in the Missouri bottoms, its
> greatest magnitude. Of one, which grew near Loutre Island,
> there had been made two hundred fence-rails, eleven feet in
> length, and from four to six inches in thickness. A cotton-tree
> [cottonwood], in the same neighborhood, produced thirty thou-
> sand shingles, as we were informed by a credible witness.

On July 3, the *Western Engineer* passed the mouth of the Gasconade
River. Independence Day passed with little celebration. An opossum
shot by Biddle, with a few glasses of wine, provided them with a meal
slightly more festive than usual. "Game is plenty," reported Peale,
"but it requires some knowledge of the country to shoot them. There
are many salt licks that are much resorted to. Wild cats (*Felis Lynx*)
are numerous. [Presumably these were bobcats (*F. rufus*), not lynx.]
Also the common Ground Squirrel (*Sciurus Striatus*) [doubtless the
eastern chipmunk, *Tamias striatus*]."

> JULY 5. TP: Took wood on board and started at 9 o'clock.
> At the plantation we were at, we procured some honey. The
> man had a number of bee hives, and has just caught another
> swarm. All that he has he had caught in the woods, where bees
> are becoming numerous. The inhabitants say that before the
> country came into the possession of Americans, there was not

a bee to be seen, and they were unknown. . . . At 10 o'clock anchored at Côte Des Sans Dessein to clean our boiler. Mr. Say and myself went up the banks of the river, chased a ground hog (*Arctomys Monax*). It ran into a hole and was soon dislodged by a stick and as it came out, I stuck it with my knife and killed it. Encamped on shore with Mr. Seymour in a tent.

The groundhog was, of course, the common woodchuck, now called *Marmota monax*. It is noteworthy that honeybees were already abundant in the wild. They were introduced from Europe at a very early date and had obviously become well established in the lower Missouri Valley by 1819. Even today, honeybees are commonly referred to simply as bees, though there are thousands of species of wild bees native to America.

Côte Sans Dessein had been founded by several families of French descent in 1808. The citizens had defended the village against an Indian attack during the War of 1812, though with some loss of life. There was now "a tavern, a store, a blacksmith's shop, and a billiard table." Near the town, Say collected a robber fly that he later described as *Laphria fulvicauda* (now *Andrenosoma fulvicauda*). This was later illustrated in color in his *American Entomology*, drawn by Titian Peale. There was an abundance of scouring rushes, or horsetails (*Equisetum*), along the river, affording "an indifferent pasturage" and sometimes proving fatal to horses.

Consumption of milk in the summer along the Missouri often produced "milk sickness," sometimes causing death. It was often ascribed to cattle feeding on poisonous plants. Baldwin believed that a form of typhus, "produced by putrid exhalations," was a more likely explanation. Quite probably, these were cases of typhoid, resulting from contamination by milk handlers who carried the disease organisms.

On July 6, the expedition passed the mouth of the Osage River, not far from the site of present-day Jefferson City. On the following day the *Western Engineer* ran aground on a sandbar and, soon after being freed, ran into a snag. As if that were not enough, one of the valves on the boilers became displaced, and the engine failed. These problems were eventually overcome, but progress continued to be slow because of the strong current and the difficulty of obtaining wood of good quality. Peale noted the abundance of chimney swifts and supposed that they nested in hollow trees, since there were few chimneys to be found. In several places there were Indian paintings on the rocks along the river, and Peale sketched some of them. The paintings included animals, humans, bows and arrows, and abstract figures.

On July 13, the *Western Engineer* reached Franklin, 200 miles above the mouth of the Missouri and the most important town west of St. Louis. The town boasted more than a hundred log houses and a few built of brick, a courthouse, a jail, and even a weekly news-paper. Although Franklin had become a major settlement, the river was already eroding the banks in front of the town and would one day spell its doom. Boonville, across the river, seemed to the natu-ralists to occupy "a more advantageous situation" and was "probably destined to rival, if not surpass, its neighbor." This proved to be true. Within a few years, in any case, Independence would replace Frank-lin as a point of departure for the West.

When Long's ship arrived, there "must have been upwards of a hundred" spectators on the bank, wrote Peale. Only once before had a steamboat ascended the Missouri this far. The ship remained in Franklin for nearly a week. This gave the passengers time to visit the brine springs known as Boone's Lick, not far away. Daniel Boone had discovered the springs on one of his wanderings, and with two of his sons had established a small industry that supplied salt by keelboat to St. Louis and other communities in the lower Missouri and Mississippi Valleys. Salt was an essential commodity, needed to cure hides and to preserve meat in days long before refrigeration was available. For Daniel Boone, central Missouri had already become overpopulated. He was quoted by the St. Louis *Enquirer* as complain-ing that "I had not been two years at the lick before a d——d Yankee came, and settled down within one hundred miles of me." He was now an old man, living with his son Nathan near St. Charles; he died the following year. But the springs were still in production.

> EJ: We visited one establishment for the manufacture of salt. The brine is taken from a spring at the surface of the earth, and is not remarkably concentrated, yielding only one bushel of salt to each four hundred and fifty gallons. Eighty bushels are manufactured daily, and require three cords of wood for the evaporation of the water. . . . The banks of the ravine in which this spring rises, still retain the traces of those numerous herds of bisons, elk, and other herbivorous animals, which formerly resorted here for their favorite condiment.

At Franklin, a local citizen reported that in 1816 he had discovered the grave of a white man, clad in an officer's uniform and left in a sitting position surrounded by a crude log enclosure. He had been scalped, and the mats beneath him were of Indian origin, suggesting that the officer had been killed by Indians but for some reason honored with an elaborate burial. A walking stick found beside him was engraved with the initials "J.M.C.," and a button had the word "Philadelphia" on it. This may or may not have been the grave of Jean Baptiste Champlain, an envoy of Manuel Lisa who had disappeared on a trading mission to the Arkansas River. Locally it was considered probable that the body was that of a Spanish officer who had been killed in a skirmish with Indians in that area in 1815. This seems a more likely explanation, though there were those who maintained that it was Champlain and that Ezekiel Williams was guilty of his murder (as discussed in Chapter 1).

Peale noted that the birds he was seeing were still mainly those of eastern distribution. Carolina wrens and parakeets were common. At Franklin, the company of naturalists was reduced by one. Baldwin had hoped that his health would improve as the expedition moved westward, but it was not to be.

EJ: Dr. Baldwin's health had so much declined that, on our arrival in Franklin, he was induced to relinquish the intention of ascending farther with the party. He was removed on shore to the house of Dr. Lowry, intending to remain there until he should recover so much strength as might enable him to return to his family. But the hopes of his friends, even for his partial recovery, were not to be realized. He lingered a few weeks after our departure, and expired on the thirty-first of August. His diary, in which the latest date is the eighth of August, only a few days previous to his death, shows with what earnestness, even in the last stages of weakness and disease, his mind was devoted to the pursuit, in which he had so nobly spent the most important part of his life.

There follows, in James's *Account*, a series of extracts from Baldwin's diary. It is primarily a list of plants observed around Franklin and will not be included here, as it reports a flora of the lower Missouri basin that was already reasonably well known. Some of the woody plants mentioned were basswood, honey locust, papaw, and smooth sumac; herbaceous plants included pokeweed, black nightshade, wild bergamot, bellflower, white avens, and diverse composites, including Canada fleabane, Indian plantain, and yarrow.

While at Franklin, the decision was made to send another group overland, to rejoin the ship at Fort Osage (some forty miles east of where Kansas City now stands). The party consisted of Say, Jessup, Seymour, Dougherty, Biddle, and two others. Peale had been suffering from a foot infection and stayed on board. The *Western Engineer* left Franklin on July 19, but that day gained only three miles against the swift current. The engine valves had become so worn from the silt in the river that they were leaking, and another day was spent making repairs. Peale, O'Fallon, and William Swift took advantage of the delay to do some hunting, but game was scarce. They did take a raccoon and a partridge.

Peale found a nest of a wood rat that contained "near a cart load" of plant material; the nest was lined with "down of the button wood ball [sycamore]." He also found a nest of "the Capped Flycatcher (*Muscicapa pusilla*)." There were two white eggs, their ends with flecks of brownish-purple. In identifying this bird, Peale was following Alexander Wilson's *American Ornithology*. Wilson had called it the "green black-capped flycatcher," though in fact it was not a flycatcher but a warbler, now called Wilson's warbler (*Wilsonia pusilla*). It is puzzling that Peale found a nest in central Missouri, well south of its normal breeding range. Had he confused it with a species of similar coloration?

The record of the trip from Franklin to Fort Osage tells mostly of navigation problems. Unfortunately Peale's diary, following the arrival at Fort Osage on August 1, has been lost, though it may have been available to James when he compiled his report on the expedition. Since Peale was not part of the overland trip to Fort Osage, we must rely on James, who doubtless received his information from Say, who was in charge of the group. The trip was less plagued by problems than the previous one. Perhaps the naturalists were learning more about survival in little known and sparsely settled country.

Say and his party traveled through forested bottomlands along the Missouri and out onto prairies "where the high grass and weeds rendered their progress difficult and laborious." There they saw sandhill cranes and prairie hens (presumably, greater prairie-chickens). They also saw four "Mississippi kites": "The forks of the tail of this bird are so much elongated as to resemble some fortuitous appendage, for which, at first sight, they are often mistaken." These were surely swallow-tailed kites, which once ranged north into Missouri but now are uncommon and restricted to the Gulf states.

At a hunter's cabin, Say "had an opportunity to examine a young black wolf, which was confined by a chain at the door of the hut. . . . When fed on meat the ferocity of his disposition manifested itself in attempts to bite the children. It was ordinarily fed on bread and milk." The hunter entertained the men with hunting tales. He claimed to have killed fifty bears and seventy deer the preceding autumn. These were white-tailed deer, and in a long footnote Say provided measurements of an individual deer shot later by the expedition, along with life history data.

Say's group reached Fort Osage well before the *Western Engineer*, and Say took occasion to study some of the fossils found in local limestone. One of the officers stationed there brought him a metallic wood-boring beetle, which he later named *Buprestis confluenta*. This is perhaps the most beautiful of North American beetles—glossy green, its wing covers speckled with yellow. It was illustrated in color in Say's *American Entomology*.

Fort Osage had been established in 1808 on a bluff overlooking the Missouri, about forty miles east of its junction with the Kansas River, guarding the western frontier of permanent settlements. There was a pentagonal stockade that enclosed buildings for living quarters, at that time housing a rifle regiment. Here at the extremity of western expansion, James was led to speculate on the "charms" of living in such remote places, free from the "uneasy restraints inseparable from a crowded population" and "dependent immediately and solely on the bounty of nature." Offsetting difficulties there surely were, but settlers the men met along the Missouri often spoke of moving still farther west whenever it became possible. James wrote of the "charms" of frontier life as he prepared the final report of the expedition, after his return east, and to a degree he may have been expressing his own vision of the frontier, for in later life he often spoke of returning west, and he eventually settled in Iowa.

The ship was delayed at Fort Osage for several days while Long waited for a break in the weather so that he could make astronomical measurements to determine the fort's exact latitude. In the meantime, he organized still another overland expedition, this one more ambitious than the previous two.

AUGUST 6. EJ: Wishing to extend our examinations between Fort Osage and the Konzas [Kansas] river, also between that river and the Platte, a party was detached from the steamboat, with instructions to cross the Konzas at the Konza village, thence to traverse the country by the nearest route to the Platte, and to descend that river to the Missouri. The party consisted of Mr. Say, to whom the command was entrusted, Messrs. Jessup, Peale, and Seymour, Cadet Swift, Mr. J. Dougherty, and five soldiers. They were furnished with three pack-horses, and a supply of provisions for ten days. Thus organized and equipped, they commenced their march on the afternoon of August 6th, accompanied by Major Biddle and his servant.

We will pick up the story of Say's group later. The *Western Engineer* proceeded upstream on August 10, its progress "much impeded by shoals and rapids." Beyond Fort Osage, the explorers passed the mouth of the Kansas River, "so filled with mud . . . as scarcely to admit the passage of our boat." Three days later, they went onward to Cow Island (Isle au Vache) (about halfway between present Atchison and Leavenworth, Kansas), the site of Camp Martin, a military post that had been established in 1818 by a company of troops that had preceded the main body of the expedition by keelboat. Since the exploring party was now well ahead of Colonel Atkinson's troops and "might be exposed to insults and depredations" from the Indians, Long requested that a small group of riflemen accompany him, by keelboat, until a winter camp was established.

At Cow Island, the men awaited the arrival of a group of Kansas Indians for a council arranged by O'Fallon. The Indians arrived on August 24; after speech-making, they examined the ship and "manifested some surprise at the operation of the steam-boat." On the following day, the augmented expedition proceeded up the Missouri.

SEPTEMBER 1. EJ: [W]e were under the necessity of remaining encamped near the mouth of Wolf river, that some repairs might be made to the steam engine. Here we sent out some persons to hunt, who after a short time returned, having taken a deer, a turkey, and three swarms of bees, which afforded us about half a barrel of honey. On the trees which margin the river, we frequently observed a fine species of squirrel, which possesses all the graceful activity of the common gray squirrel, as it leaps from bough to bough.

Examples of these squirrels were collected, and a description was included in the *Account* by Thomas Say. He named the species *Sciurus macrurus* (Greek for "big tail"), noting that it "seems to approach the *Sc. rufiventer*," which Geoffrey St. Hilaire had described in 1803. When John James Audubon and John Bachman published their *Viviparous Quadrupeds of North America* in 1851, they realized that the name *macrurus* had been used earlier for another kind of squirrel, so they renamed it in honor of Thomas Say, *S. sayii*. However, neither name is now used for these squirrels, since they are considered identical to St. Hilaire's, whose name is used, following the law of priority. This is the midwestern race of the fox squirrel, now called *S. niger rufiventre*.

Near the Wolf River, the ship was hailed from the shore by Dougherty, who had been with Say's party. He reported that most of the group was a short distance downstream, but Say and Jessup had taken sick and had stayed behind with the troops at Cow Island. Since the overland party had planned to meet the rest of the expedition at the mouth of the Platte, this came as a surprise, but at least all were safe. After Dougherty and his companions were aboard, the ship proceeded upstream.

The overland expedition had been out for nearly four weeks and had gone as far as the junction of the Big Blue River with the Kansas River, where the village of the Kansas Indians was located. This is near the site of present-day Manhattan, Kansas. From there, they had traveled northeast to Cow Island, without coming anywhere near the Platte, as originally planned.

After crossing the plains for some distance, Say's party saw ravens for the first time and killed several rattlesnakes. Blowflies were abundant, "attacking not only the provisions of the party, but depositing their eggs upon the blankets, clothing, and even on the furniture of the horses." Following the course of the Kansas River, they suffered from heat and the roughness of the terrain. Several in the party had dysentery. By August 16, they were approximately at the site of Topeka. Four days later, they saw the Indian village in the distance, and they stopped "to arrange their dress, and inspect their firearms." In the village they were received "with the utmost cordiality."

Say described the village of the Kansas Indians as consisting of about 120 lodges clustered irregularly on a plain a short distance from the river. Each lodge was built in a shallow, circular depression and had a roof supported by vertical posts and covered with smaller branches over which were laid mats of grass or bark; the whole was then covered with earth, leaving a hole in the middle for the escape of smoke. The interior of each lodge had a central fireplace and walls lined with mats and with simple bedsteads covered with bison robes. Some conception of the interior may be obtained from Seymour's painting of a Kansas war dance; the outside, from a sketch by Peale. These fixed lodges were very different from the movable skin tipis of plains Indians they would later observe and sketch.

AUG. 20. TS: After the ceremony of smoking . . . the object which the party had in view in passing through their territories was explained to them, and seemed to be perfectly satisfactory. At the lodge of the principal chief they were regaled with jerked bison meat and boiled corn, and were afterward invited to six feasts in immediate succession.

They commonly placed before us a sort of soup, composed of maize of the present season . . . boiled in water, and enriched with a few slices of bison meat, grease, and some beans, and to suit our palates, it was generally seasoned with rock salt, which is procured near the Arkansa river.

This mixture constituted an agreeable food; it was served up to us in large wooden bowls, which were placed on bison

Titian Peale, sketch of lodges of the Kansas Indians, mid-August, 1819. (From the sketchbooks of Titian Ramsay Peale, Yale University Art Gallery, gift of Ramsay MacMullen, M.A.H. 1967)

robes or mats, on the ground; as many of us as could conveniently eat from one bowl sat round it, each in as easy a position as he could contrive, and in common we partook of its contents by means of large spoons made of bison horn. We were sometimes supplied with uncooked dried meat of the bison, also a very agreeable food, and to our taste and reminiscence, far preferable to the flesh of the domestic ox.

Other foods consisted of beans, pumpkins, melons, and corn prepared in several different ways—all of these grown by the women. After "six feasts in immediate succession," it is little wonder that Say and Jessup became ill after a few days!

Say described the dress and customs of the Kansas Indians in some detail, but most of that information will be omitted here. He found the women to be "small and homely, with broad faces." The men scrupulously removed all their hair except for a patch on the back of their head "to supply their enemy with a scalp, in case they

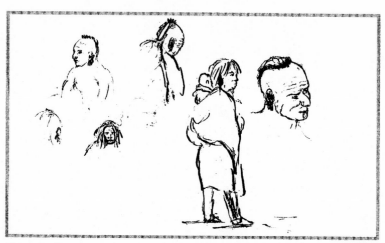

Titian Peale, portraits of Kansas Indians. (From the sketchbooks of Titian Ramsay Peale, Yale University Art Gallery, gift of Ramsay MacMullen, M.A.H. 1967)

should be vanquished." This strip of hair was decorated with an eagle feather or a deer tail. The bodies of both men and women were usually tattooed.

While the data gathered by Say proved valuable to ethnologists, he had by no means abandoned entomology. Near the village, he collected two new species of cicadas, one a handsome species that produced a particularly harsh call. He named it *Cicada dorsata* (now *Tibicen dorsata*). In the course of the expedition, Say was to discover and name five species of cicadas. These large insects (often incorrectly called locusts) drone from trees and bushes on summer days in many parts of the country, each species producing its own characteristic "song."

At the Kansas village, Say also collected a darkling beetle that he named *Blaps suturalis* (now *Eleodes suturalis*). His comments following the description of this species (in *American Entomology*) are worth quoting.

Titian Peale, watercolor of cicadas. (American Philosophical Society)

TS: Whilst sitting in the large earth-covered dwelling of the principal chief, in presence of several hundred of his people, assembled to view the arms, equipments, and appearance of the party, I enjoyed the additional gratification to see an individual of the fine species of *Blaps* running towards us from the feet of the crowd. The act of empaling this unlucky fugitive at once conferred upon me the respectful and mystic title of "medicine man," from the superstitious faith of that simple people.

On August 23, Say's party purchased bison meat, pounded corn, moccasins, and other supplies, and headed north toward the Platte with three Indians as guides. At their first camp, seven miles from the village, Dougherty and one of the Indians went out hunting. While they were away, a war party of over a hundred Pawnees approached the camp. The Pawnees were then at war with the Kansas, and the two Kansas Indians in camp quickly disappeared into the bushes, leaving Say and his small band at the mercy of the Pawnees. Although fully armed and painted for war, the Indians entered the camp peacefully. But they soon began to steal food, blankets, firearms, and horses. The warriors drew their arrows and cocked their guns, but they retreated without doing injury just as Dougherty returned.

The men could scarcely proceed without horses or supplies, so they sent an envoy to the Kansas village and then returned there for a day. In the evening, they were treated to a war dance, accompanied by shouts, drums, and rattles. The rattles consisted of "strings of deer's hoofs, some part of the intestines of an animal inflated, and enclosing a few small stones."

Engraving in the *Account*, after Samuel Seymour, watercolor of a war dance in the interior of a lodge of the Kansas Indians.

Since their supplies had been lost, the men had no means of bartering for horses. However, a Frenchman who had been living with the Indians agreed to accompany them and to supply two packhorses and a saddle horse for Say, who was ill. There was no hope of completing their trip as planned, so they set off directly for Cow Island, where they might intercept the ship. They camped for one night on Grasshopper Creek, where they saw several "orbicular lizards" (horned lizards, *Phrynosoma douglassii*). Arriving at Cow Island too late to meet the ship, most of the party "by great exertion" boarded the ship upstream at Wolf River, leaving Say and Jessup to recover at Cow Island with the troops who were stationed there.

James's *Account* includes several remarks on the country in the vicinity of the village of the Kansas Indians, doubtless based on Say's impressions. As one approaches "the borders of that great Sandy Desert, which stretches eastward from the base of the Rocky Mountains," one finds cacti and other plants that "delight in a thirsty muriatiferous [alkaline] soil." Although woodlands are now confined to stream valleys, says James, forests may some day be planted on the plains, and "wells may be made to supply the deficiency of running water." Thus to a degree he foresaw the time when the Great Plains would be watered from aquifers, though not for the growing of forests. Among the trees listed as occurring along the streams were honey locust, ash, walnut, and Kentucky coffee tree (*Gymnocladus dioica*). The seeds contained in the large, woody pods of this tree were often used as a substitute for coffee on the frontier.

Proceeding north along the Missouri, the passengers on the *Western Engineer* saw several deer swimming across the river, and shot several from the ship. They also found a hive of bees, "and the honey they afforded made a valuable addition to our provisions, consisting now in great measure of hunters' fare." They caught several catfish, "some of them weighing more than fifty pounds."

EJ: We have seen . . . a pretty species of sparrow, which is altogether new to us; and several specimens of a serpent have occurred, which has considerable affinity with the pine-snake of the southern states, or bull-snake of Bartram.

It was Say's practice to include descriptions of birds, reptiles, and mammals in the expedition's *Account*, while saving the insects for inclusion in his *American Entomology* or other publications. Thus he provided descriptions of both the sparrow and the snake in lengthy footnotes. The sparrow he called *Fringilla grammaca* (now *Chondestes grammacus*). This is the lark sparrow, one of the handsomest of the sparrow clan. "They run upon the ground like a lark, seldom fly into a tree, and sing sweetly," reported Say. Say named the snake *Coluber obsoletus*. This is the pilot black snake, now called *Elaphe obsoleta*. This large, nonpoisonous snake kills its prey by constriction. The name "pilot" apparently arose through a superstition that these

Titian Peale, pencil-and-watercolor sketch of a lark sparrow. June 23, 1819. (American Philosophical Society)

snakes lead other snakes to safety when danger threatens. The naturalists took these snakes near the western extremity of their range.

Somewhere on the Missouri, no one is quite sure where, Say found and described a beetle that he named *Doryphora 10-lineata*. This is the famous Colorado potato beetle, now called *Leptinotarsa decemlineata*. The original host plant is believed to have been a wild member of the potato family. Once the settlers began to plant potatos in the West, the beetles found themselves with an endless new food supply, and they swept across the country and even into Europe, leaving behind devastated potato fields and at times near-starvation in the populace. Of all the hundreds of insects Say described, this is without doubt the most notorious.

Say's name *10-lineata* (ten-striped) describes these beetles well, but there seems no justification for associating the state of Colorado with the vernacular name, which was applied much later. Say stated, in an article published in 1824, that the species "inhabits Missouri and Arkansa" and was "not uncommon on the Upper Missouri." Specimens collected "on the Arkansa[s]" were assigned to a variety of slightly different color. While the latter may have been taken in what is now the state of Colorado, there is no doubt that Say associated the name primarily with specimens from the Missouri Valley.

On September 15, the expedition arrived at the mouth of the Platte, appropriately named the "flat" river, since its mouth "exhibited a great extent of naked sand-bars, the water . . . flowing almost unseen through a number of small channels." They were to see much of the Platte the following year.

> SEPTEMBER 15. EJ: Above the Platte, the scenery of the Missouri becomes much more interesting. The bluffs on each side are more elevated and abrupt, and being absolutely naked, rising into conic points, split by innumerable ravines, they have an imposing resemblance to groups of high granitic mountains, seen at a distance. The forests within the valley are of small extent, interspersed with wide meadows covered with [various sedges and grasses], sometimes sinking into marshes occupied by sagittarias [arrowhead], alismas [water plantain], and others. . . . The woodlands here, as on the whole of the Missouri below, are filled with great numbers of pea vines, which afford an excellent pasturage for horses and cattle. The roots of the apios tuberosa [now *Apios americana*, groundnut or wild bean] were much sought after, and eaten by the soldiers, who accompanied us in our ascent. They are little tubers about half an inch in diameter, and when boiled are very agreeable to the taste.

On September 17, the expedition reached the trading post of the Missouri Fur Company, Fort Lisa. On the west bank of the Missouri, about half a mile above Fort Lisa and five miles below Council Bluff, where Colonel Henry Atkinson's troops were to spend the winter in "Camp Missouri," Long decided to establish his winter camp. Council Bluff had been named by Lewis and Clark, who had held an

Titian Peale, watercolor of the camp at Engineer Cantonment, February 1820. The *Western Engineer* is in the left foreground; keelboats are on the right. (American Philosophical Society)

important meeting with the Oto and Missouri Indians there in 1804; it was several miles above the location of the modern city of Council Bluffs. Within a few days, Long and his men "had made great progress in cutting timber, quarrying stone, and other preparations for the construction of quarters."

EJ: [At the site selected] a very narrow plain or beach, closely covered with trees, intervenes between the immediate bank of the river and the bluffs, which rise near two hundred feet, but are so gradually sloped as to be ascended without great difficulty, and are also covered with trees. This spot presented numerous advantages for the cantonment of a small party like ours. Here were abundant supplies of wood and stone, immediately on the spot where we wished to erect our cabins, and the situation was sheltered by the high bluffs from the northwest winds. The place was called Engineer Cantonment.

On September 20, Say and Jessup—both "nearly recovered" in health—arrived with a group of military personnel from Cow Island. A few weeks later, Long departed for Washington to be married and to await further orders, and Jessup, deciding that he had had enough privation, left permanently with him. Major Biddle had now left the exploring party to join Colonel Atkinson's military contingent; he had never seen eye to eye with Long. With the earlier loss of Baldwin, the scientific staff was reduced to three (Say, Peale, and Seymour), and the officers to two (Graham and Swift). Fortunately John Dougherty remained with the group and was of inestimable value in gathering data on the local Native Americans. Despite the diminished size of the party, and their problems in keeping warm and well fed through a long winter, much was accomplished. For once, they did not have to push on day by day, and their notebooks became crammed with information.

The *Western Engineer* remained at Engineer Cantonment through the winter, serving as a storehouse for the expedition's books and instruments. Peale sketched the ship and the cliffs behind from the camp, and after returning to Philadelphia prepared a watercolor that Kenneth Haltman has termed "perhaps the most literal, visually persuasive assertion of white cultural control over the western landscape in Peale's expedition artwork." Haltman feels that by portraying the ship puffing smoke, displaying its flag, and bearing US conspicuously on its wheel, the young Titian Peale attempted to symbolize his country's conquest of the West: "The steamboat was an eastern dynamo sent west, burning its way into the wilderness, devouring trees as fuel and converting nature into European-American culture . . . an intellectual machine determined to overcome any resistance in ingesting and processing Nature (which, of course, in the last century *included* Indians)." But in a few months, the *Western Engineer* would be left behind as the men penetrated the wilderness much more deeply.

Five

OVERWINTERING AT ENGINEER CANTONMENT

ON OCTOBER 11, LONG AND JESSUP began their trip east by starting downstream in a canoe. Lieutenant James Graham was left in charge of the *Western Engineer*, with instructions to do certain repairs and to run the engines from time to time. He was also to make "celestial and barometric observations," keep weather records, and measure the height of surrounding highlands. William Swift was to assist him. In fact, the *Western Engineer* had reached the end of its trip up the Missouri, though that may not have been obvious to Long at the time. Nor would Colonel Henry Atkinson's troops ascend the river farther at this time, in steamboats or otherwise. The *Western Engineer* was the only steamboat to reach Council Bluff, but it had taken it nearly three months to cover the 600 miles from St. Louis, averaging only about 6 miles a day. Long was to receive considerable criticism on his arrival in Washington, from both Congress and the press. Indeed, the entire Yellowstone Expedition had proved a disappointment.

Before leaving, Long left instructions for the small group of naturalists remaining.

OVERWINTERING AT ENGINEER CANTONMENT

SL: Mr. Say will have every facility afforded him that circumstances will admit to examine the country, visit the neighbouring Indians, procure animals, &c. for the attainment of which he will call on Lt. Graham, who is authorized to make the expenditures in behalf of the expedition that may be deemed reasonable and necessary, and afford all aid in his power, consistent with the performance of other duties. Mr. Seymour or Mr. Peale will accompany him, whenever their services are deemed requisite.

Major O'Fallon has given permission to Mr. Dougherty to aid the gentlemen of the party, in acquiring information concerning the Indians, &c.; this gentleman will, therefore, be consulted in relation to visits, and all kinds of intercourse with the Indians, that may be necessary in the prosecution of the duties of the expedition.

It is believed, that the field for observation and inquiry is here so extensive, that all the gentlemen of the expedition will find ample range for the exercise of their talents in their respective pursuits; and it is hoped, that through their unremitted exertions and perseverance, a rich harvest of useful intelligence will be acquired.

TS: The leisure we enjoyed after our arrival at Engineer Cantonment, afforded the opportunity of making numerous excursions to collect animals, and to explore the neighbouring country.

Say was soon out collecting insects. On a sandbar near the camp, he found great numbers of a small beetle flying in the evening. They belonged to a group now called variegated mud-loving beetles. He later described the species as *Heterocerus pallidus*. In a quarry from which building stone had been taken, Say found a new species of bombardier beetle, which he named *Brachinus cyanipennis*. These beetles are noteworthy for their ability, when disturbed, to discharge

Colorado potato beetles. Thomas Say collected specimens of this notorious pest on the Missouri in 1819.

Boxelder bug. This common insect was described by Say from specimens collected at Engineer Cantonment.

hot, caustic fluids with an audible "pop." Perusal of Say's publications reveals that he described at least twelve new species of beetles and at least six new species of true bugs (Hemiptera) from Engineer Cantonment. One of these, the box elder bug (*Leptocoris trivittatus*) is all too well known to homeowners who live anywhere near box elder trees, since the bugs swarm into houses during the fall and winter. They are, however, harmless to people and quite beautifully adorned with red stripes.

The stone quarry contained "many large fissures, in which were found a number of serpents that had entered there for the purpose of hybernating. Of these, three species appear to be new." One of these was the blue racer, a slender, elegant snake whose tendency to climb over the tops of bushes may have given rise to the "hoop snake" myth. The other two were the red-sided garter snake and the western ribbon snake. It is not unusual for snakes of more than one species to hibernate together. Say's scientific names are now used for subspecies of relatively widely distributed snakes.

Cliffs behind the camp were of sandstone and limestone and were rich in fossils. There were sharks' teeth, crinoids, and an abundance of shells of mollusks. In a long footnote, Say described many of these.

In a pit trap dug by Peale for catching a wolf, Say found and described a new shrew, which he named *Sorex parvus* (now *Cryptotis parva*, least shrew [a mere 2⅜ inches from tip of nose to root of tail]). A second species he called *Sorex brevicauda* (now *Blarina brevicauda*, short-tailed shrew). Both, we now know, are species widely distributed in the eastern states, but at Council Bluff were near the western extremities of their ranges. Had they not fallen into the trap, Say may never have found them, as shrews live in shallow runways in the soil or litter and are rarely observed. They are voracious predators, often consuming twice their own weight in insects and worms every day. Both of Say's species names for the shrews remain valid, but he was less fortunate in the two kinds of bats he described. These were the hoary bat and the big brown bat. Both had been named and described two decades earlier by French biologist Ambroise Palisot de Beauvois.

Prairie wolves (coyotes) barked and howled around the camp at night, and Say came to admire their "wonderful intelligence." Peale tried to catch one alive with a baited, inverted box supported by stakes. The coyote escaped with the bait. Several other kinds of traps were equally unsuccessful. Finally, a log supported by two sticks, one of them baited, succeeded in taking one, but it was killed. Say wrote that "this animal, which does not seem to be known to naturalists . . . is most probably the origin of the domestic dog, so common in the villages of the Indians of this region, some of the varieties of which still retain much of the habit and manners of this species."

The coyote was unknown to naturalists in the sense that it had never been formally described and given a scientific name, but it was familiar to many who had traveled or settled in the West. Lewis and

Clark saw and heard coyotes many times and even sent a coyote skeleton to Thomas Jefferson. But it remained for Say to describe it in detail and provide a name, *Canis latrans* (*latrans* being the Latin word for "howling").

Lewis and Clark were also all too familiar with gray wolves, and often had to hang their meat out of reach of them. Like Lewis and Clark and others, Say recognized the wolf as different from the coyote: "The aspect of this animal is far more fierce and formidable than . . . the prairie wolf, and is of more robust form. . . . It diffuses a strong and disagreeable odour, which scented the clothing of Messrs. Peale and Dougherty, who transported the animal several miles from where they killed it on the cantonment." Say provided a formal description of the gray wolf, calling it *Canis nubilus*. His name is retained for the extinct midwestern race of the wolf, *C. lupus nubilus*. The wolf ranges (or used to range) throughout Eurasia as well as North America, and has long played a role in fiction and folklore. It was the Swedish naturalist Linnaeus who formally described *C. lupus* (*lupus* being the Latin word for "wolf").

Wolves and coyotes were to haunt the expedition's campsites throughout its trek through the West. Nowadays, we may read about wolves in books or listen to their howls on records, though a few packs live a precarious life in the northern Rockies. The more adaptable and less threatening coyotes, meanwhile, have expanded their range into the eastern states and even into the suburbs of major cities. To moderns, the nocturnal chorus of a pack of coyotes serves to remind us that a bit of nature remains unconquered. But Long and his men soon tired of wild canines.

The food of the men at Engineer Cantonment was constantly augmented by hunting. Peale on one occasion "killed two deer at a single shot and with one ball." Another time Peale and Dougherty returned from a hunt "having killed twelve bisons out of a herd of several hundreds they met with near Sioux river, and brought us a seasonable supply of meat." Elk were also sometimes on the menu, and even a skunk provided "a remarkably rich and delicate food."

Members of the party also fished through the ice of a pond, obtaining "one otter and a number of small fishes, amongst which three species appeared to be new." There was a good supply of flour, from which the cook prepared "bread fully equal, in point of excellence, to any we have ever eaten." There was no coffee, but the fruit of the Kentucky coffee tree was found to provide "a palatable and wholesome beverage."

On December 5, the "gentlemen of the party" were invited to dine with Manuel Lisa. Present at the banquet was Lisa's wife, Mary, the first white woman to have ascended the Missouri this far. She was a handsome woman and must have charmed the explorers, now so far from settlements. She surely intrigued the Indians, who followed her about everywhere. The Indians provided a feast for her, serving their most elegant repast: dog. Somehow, Mary (who had been brought up in Connecticut) managed to pretend to eat, while depositing the meat in a handbag she held on her lap.

It is said that Mary Lisa did not speak Spanish or French, and Manuel Lisa barely spoke English. According to one story, "this was why they were able to live so happily together." But their marriage was short, from 1818 to Lisa's untimely death in 1820. Manuel also had an Indian wife, of the Omaha tribe, by whom he had two children. It was common for traders to secure their relationships with the Indians by taking a wife from among them. It is likely that Mary Lisa knew or at least learned of this Indian wife, since Manuel provided for his part-Indian children in his will. Mary lived to be eighty-seven, dying in 1869.

In midwinter, there were fewer opportunities for natural history, but the group at Engineer Cantonment took advantage of the time to study the customs of the local Indians. A committee of members of the American Philosophical Society had provided instructions to the expedition as to the kinds of information desired. Since several members of the committee were physicians, they especially wanted to know what diseases were prevalent among the Indians and what remedies they used. Say, through Dougherty, obtained in-

Engraving after Samuel Seymour, watercolor of an Oto council. (Academy of Natural Sciences of Philadelphia)

formation on these matters from members of several tribes, and he had long conversations with Dougherty himself: "This gentleman with great patience, and in the most obliging manner, answered all the questions which I proposed to him, relating to such points in their manners, habits, opinions, and history, as we had no opportunity of observing ourselves."

Well over a hundred pages of James's *Account* consists of Say's somewhat rambling report on the Indians. Seymour and Peale often sketched the Indians' hunting forays, tipis, dances, and ceremonies. Some of their illustrations appeared in the final report of the expedition, published in 1823. So their drawings anticipated by a decade George Catlin's and Karl Bodmer's more celebrated paintings of Indian life.

Major Benjamin O'Fallon frequently summoned tribes for a council at Camp Missouri. The first to arrive, in early October, was a group of Otos, who honored the men of the expedition with a dance accompanied by "rude instrumental and vocal music." One of the chief warriors, Ietan (or Chon-Mon-I-Case) narrated his exploits at length. Ietan had stolen horses from several tribes, struck the bodies of Pawnee and Sioux warriors, and even participated in an attack on a Spanish camp. In 1821, Ietan was to be part of a group of Indians taken to Washington by O'Fallon. In Washington, his portrait was painted by Charles Bird King as he wore a headress crowned by bison horns and a necklace of grizzly-bear claws.

When the Pawnees arrived, O'Fallon "addressed them in a very austere tone and manner," whereupon they returned much of the property they had stolen near the Kansas village. In November, a band of Sioux arrived and were invited to visit the *Western Engineer.* They were hesitant, but once aboard they "appeared much delighted." They were shown some of the workings of the ship and were astonished at the operation of the howitzers.

There were opportunities to visit villages of the Omaha, Oto, Missouri, and Iowa Indians. Like other plains Indians, these tribes depended heavily on bison both for food and for robes and skins for clothing and shelter. They also grew corn and pumpkins, which they prepared in various ways. The squaws collected roots of groundnut (*Apios americana*) and the seeds and roots of lotus, or pond nut (*Nelumbo lutea*).

Say noted that the women searched for and ate the lice found in one anothers' hair: "One of them, who was engaged in combing the head of a white man, was asked why she did not eat the vermin; she replied, that 'white men's lice are not good.' " Say also tells of a custom among the Snake Indians of collecting ants from their mounds and placing them in a bag that is taken to a stream to wash out the dirt and sticks. "The ants are then placed upon a flat stone, and by the pressure of a rolling-pin, are crushed together into a dense mass, and rolled out like pastry. Of this substance a soup is prepared, which is relished by the Indians, but is not at all to the taste of white men."

It would be beside the point of this narrative to review Say's observations in detail, so I shall select a few points bearing on the Indians' use of native plants and animals. They made spoons of bison horns, often ornamented, and a digging instrument from the scapula of a bison when they could not obtain tools from traders. The strings of their bows were made of twisted bison sinews. The bows were made of osage orange (*Maclura pomifera*) or hop hornbeam (*Ostrya virginica*). Arrows were made of arrowwood (*Viburnum*) and were fitted with turkey feathers to stabilize their flight. Feathers of turkeys, eagles, and other large birds were used in ceremonial clothing. For smoking, they often mixed tobacco (when available from traders) with dried leaves of smooth sumac (*Rhus glabra*) or silky dogwood (*Cornus Amomum*). When none of these were available, the inner bark of *Viburnum* could be substituted.

After the hunters had killed a bison (usually a cow), they wasted very little, only the feet and most of the head. After the hide was removed, the carcass was cut in pieces and carried back to camp by the squaws, who prepared the meat for storage and for immediate consumption.

> TS: The vertebrae are comminuted by means of stone-axes . . . ; the fragments are then boiled, and the rich fat or medulla which rises, is carefully skimmed off and put in bladders for future use. The muscular coating of the stomach is dried; the smaller intestines are cleaned and inverted, so as to include the fat that had covered their exterior surface, and then dried; the larger intestines, after being cleaned, are stuffed with meat, and cooked for present eating.
>
> The meat, with the exception of the shoulders, or hump, as it is called, is then dissected with much skill into large thin slices, and dried in the sun, or jerked over a slow fire on a low scaffold.
>
> The bones of the thighs, to which a small quantity of flesh is left adhering, are placed before the fire until the meat is sufficiently roasted, when they are broken, and the meat and marrow afford a most delicious repast. These, together with the tongue and hump, are esteemed the best parts of the animals.
>
> The meat, in its dried state, is closely condensed together

Titian Peale, watercolor of Otoes. (American Philosophical Society)

into quadrangular packages, each of a suitable size, to attach conveniently to one side of the packsaddle of a horse. The dried intestines are interwoven together into the form of mats, and tied up in packages of the same form and size.

The brains and livers of the bison were used in preparing the hides after they had been stretched, dried, and cleaned. The skins of elk, deer, and pronghorn antelope were dressed in a similar manner before being made into clothing, moccasins, or coverings for lodges. Bison robes, moccasins, and beaver skins were traded to the whites for guns, powder, vermilion, knives, kettles, mirrors, and a variety of other items.

Say reported that there were a good many "illicit amours" among the Indians, and the women sometimes practiced abortion. Considering the current controversy on this subject in our society, it may be worth quoting Say's remarks.

TS: Abortion is effected, agreeably to the assertions of the squaws, by blows with the clenched hand, applied upon the abdomen, or by repeated and violent pressure upon that part, or by rolling on the stump of a tree, or other hard body.

Say noted that the squaws sometimes experienced difficulties at the birth of their first child, in which case "the young wife calls in some friendly matron to assist."

TS: The aid which these temporary midwives afford, seems to be limited to the practice of tying a belt firmly about the waist of the patient, and shaking her, generally in a vertical direction, with considerable violence. In order to facilitate the birth, a vegetable decoction is sometimes administered; and the rattle of the rattle-snake is also given with, it is said, considerable effect. The singular appendages of this animal are bruised by pounding, or comminuted by friction between the hand, mixed with warm water; and about the quantity of two segments constitutes a dose.

As Roger Nichols and Patrick Halley point out in their book *Stephen Long and American Frontier Exploration*, some of the information gathered by the exploring party was well known to persons who had traveled or lived in the West, but there had been few attempts to describe Indian customs in such detail and to make the information available to readers in the East. They remark that much of the discussion of the Kansas and Omaha tribes, in particular, "provided an authoritative source of ethnological findings unrivaled for decades."

The winter was long and cold. On December 10, the temperature was below zero for most of the day, and on February 9, 1820, the ice on the Missouri River was sixteen inches thick. But the ice began to break up in March, and "great flights of geese, swans, ducks, brant, and cranes" were seen flying north.

The breakup of the ice made it possible to remove the sick from

Camp Missouri and send them downriver to Fort Osage for recovery. More than 300 soldiers had become ill with scurvy, and nearly 100 had died.

TS: Individuals who are seized rarely recover . . . they have no vegetables, fresh meat, nor antiscorbutics, so that patients grow daily worse, and entering the hospital is considered by them as a certain passport to the grave. . . . The causes which have been productive of all this disease, are not distinctly known. . . . But it was generally remarked that the hunters, who were much employed in their avocation, and almost constantly absent from Camp Missouri, escaped the malady.

Evidently, the men at Engineer Cantonment had a sufficiently diversified diet that they were little troubled by scurvy.

APRIL 13. TS: [This] morning we were awakened by the loud cries of the sandhill crane, performing evolutions in the air, high over their feeding grounds. This stately bird is known . . . [as] grus canadensis. It . . . is very distinct from the grus Americanus of authors, or hooping crane, although many persons have supposed it to be no other than the young of that gigantic species. The sandhill crane, in the spring of the year, removes the surface of the soil by scratching with its feet, in search of the radical tubers of the pea vine, which seem to afford them a very palatable food.

This crane is a social bird, sometimes assembling together in considerable flocks. They were now in great numbers, soaring aloft in the air, flying with an irregular kind of gyratory motion, each individual describing a large circle in the air independently of his associates, and uttering loud, dissonant, and repeated cries. They sometimes continue thus to wing their flight upwards, gradually receding from the earth, until they become

Titian Peale, watercolor of sandhill cranes, Engineer Cantonment, March 1820. (American Philosophical Society)

mere specks upon the sight, and finally altogether disappear, leaving only the discordant music of their concert to fall faintly upon the ear.

On April 20, O'Fallon led a party westward from the Missouri to visit the Pawnee villages along the shallow, sprawling river that French traders had called the Platte, and the Oto Indians, the Nebraska ("flat water"). With him were Dougherty, Graham, Say, and officers and troops from Camp Missouri. The troops entered the village to the accompaniment of bugles, fifes, and drums, much to the delight of the children. They were approached by a great number of mounted Indians, from whom there emerged a smaller group, slowly at first and then at a full gallop. Say marveled, "It is impossible by description to do justice to the scene of savage magnificence that was now displayed. Between three and four hundred mounted Indians, dressed in their richest habiliments of war, were rushing around us in every direction, with streaming feathers, war weapons, and with loud shouts and yells."

The Pawnees were led by their chief, Latelesha, who had visited William Clark in St. Louis, and by his son, Petalesharoo, a strikingly handsome warrior who the next year was to visit Philadelphia and Washington, where his portrait was painted as he wore his headress of eagle feathers and ermine tails. The portrait came to be accepted as that of a typical Indian chief for years to come, and Petalesharoo became the model of many an Indian chief in tales about the frontier, including that of Hard-Heart in James Fenimore Cooper's *The Prairie* (1827).

Much of the background for *The Prairie* was in fact based on Cooper's reading of the narrative of the Long Expedition, particularly those parts dealing with the Pawnees. In some cases, Cooper adopted his figures of speech directly from the *Account*—for example, the prairie as an ocean with its waters heaving restlessly. Cooper had never been west, but he met Petalesharoo in Washington in 1821. Hard-Heart, he wrote in *The Prairie*, "was in every particular a warrior of fine stature and admirable proportions. . . . [H]is countenance appeared in all the gravity, the dignity and it may be added in the terror, of his profession." Dr. Obed Battius, Cooper's wild-eyed naturalist who applied Latin names to everything he saw, may well have been suggested by his reading of the exploits of Thomas Say or Thomas Nuttall.

At the Pawnee villages, O'Fallon and his men were treated to a feast and dances, the latter accompanied by drums and other simple instruments. Petalesharoo suggested that his visitors play their own music to accompany the dances; but when the bugles sounded, the dancers were thrown into confusion, to the amusement of all. After discussions and an exchange of gifts, the visitors returned to their camps along the Missouri.

At Engineer Cantonment, Say and Peale faithfully recorded the arrival date of each migrating bird, and these data were included in the report of the expedition. They also kept a list of all the animals seen or taken. Their list included 10 kinds of reptiles, 11 amphibians, 34 mammals, and 144 birds (this was long before there were any handy "field guides"!). Many of the species on their list now go by different scientific or common names, but most are recognizable. The "hooping crane" is listed, as well as the bald eagle—the former now emerging (one hopes) from the brink of extinction, the latter now recovering from near extinction in major parts of its range. The passenger pigeon is listed, though not mentioned elsewhere in the expedition's reports; of course, there was no way of knowing that these common birds would, in a few decades, disappear completely from the fauna. When John Bradbury had passed through this country nine years earlier, he had been much impressed by the abundance of the pigeons: "This species of pigeon associates in prodigious flocks; one of these flocks, when on the ground, will cover an area of several acres in extent, and the birds are so close to each other that the ground can scarcely be seen." Bradbury shot 271 in a few hours before he finally "desisted."

Among the migrating birds, Say recognized several that were new to science, and he duly provided names and descriptions. One was a warbler that he called *Sylvia celatus* (now *Vermivora celatus*, orange-crowned warbler). Another he named *Limosa scolopacea* (now *Limnodromus scolopaceus*, long-billed dowitcher [a handsome sandpiper that breeds north of the Arctic Circle]). He also described the pectoral sandpiper, but that species had been described a few years earlier by Louis Vieillot from Paraguay, where the birds spend the winter. Say also described another warbler as new, calling it *Sylvia bifasciata*, but that species had been described a few years earlier by Alexander Wilson as *S. cerulea* (now *Dendroica cerulea*, cerulean warbler). Following the rules of priority, the earlier names are those now used.

Major Long, accompanied by two new members of the expedition, had left for the West in March, arriving at Engineer Cantonment on May 27. On June 1, he issued some new and quite different orders that would take the expedition into unexplored regions that might provide unlimited opportunities for the naturalists.

$\mathcal{S}ix$

NEW PLANS AND A NEW CAST
OF CHARACTERS

WHILE IN WASHINGTON, MAJOR LONG HAD reported to Secretary of War Calhoun and attempted to obtain financial support for another year of exploration. He found that Congress, irked by the slow progress and modest accomplishments of both the military and scientific contingents, had launched an investigation of the expedition. Furthermore, the Panic of 1819 had caused deep cuts in the military budget, and Calhoun's resources were limited. After much discussion, it was decided that Long would lead a low-budget, four-month trip west by land, sending the *Western Engineer* down the Mississippi to pick up the explorers on their return. The grandiose plans of the Yellowstone Expedition had been scuttled. According to Hiram Chittenden, a harsh later critic of the expedition, "[A]s a half-hearted apology to the public for its failure, a small side show was organized for the season of 1820 in the form of an expedition to the Rocky Mountains." The "side show" would also have its critics, including Chittenden, but it was to yield results far greater than might have been expected from the meager support provided by the government.

Long was permitted to recruit two persons to replace the three who had left the expedition. The man he chose to replace Thomas Biddle as journalist was Captain John R. Bell, a graduate of the United States Military Academy at West Point who had served in the War of 1812. After the war, he had been stationed for a time in Boston, and then became an instructor at West Point. He evidently did not enjoy teaching and applied for an appointment to the expedition. As a journalist, he was to prove much more faithful than Biddle. However, as noted earlier, his journals were not made public until 1957, 134 years after the publication of Edwin James's *Account*. James made no use of Bell's journals in preparing his formal report on the expedition.

Edwin James joined the expedition to replace William Baldwin and Augustus Jessup. He came from a Vermont family and was a graduate of Middlebury College. He studied medicine for three years under his two brothers, who were physicians in Albany, New York. He also studied botany informally under Amos Eaton and John Torrey, both distinguished botanists. He was interested in geology and had joined the American Geological Society, which had been founded in 1819. By 1820, he had already published articles on botany and geology. Although he was only twenty-three years old, James was uniquely qualified to fill the roles of physician, botanist, and geologist.

Long, Bell, and James left Pittsburgh by steamboat on March 31, arriving in St. Louis three weeks later. Bell was appalled by the lack of cleanliness in the streets and by the homes in St. Louis: "Unless more attention is paid to it by the inhabitants and authorities of the town, it is doomed to be a very sickly place . . . and many a new inhabitant and stranger will have arrived here, to close the scene of his earthly career." Bell, James, and Long visited Governor William Clark, who showed them the rifle and other equipment that he had carried on his trip to the Pacific and back. Bell studied them with "veneration," since he was soon to undertake a similar expedition and "might not live to return and enjoy the pleasure of exhibiting to my friends my equipments."

Long expected that funds to finance the summer's explorations

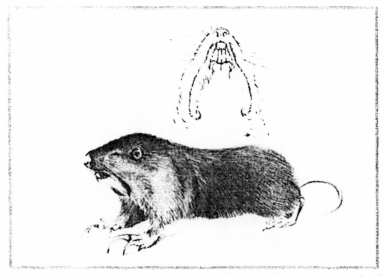

Titian Peale, pencil-and-watercolor sketch of a pocket gopher, with details of the jaws, July 21. 1819. (American Philosophical Society)

would precede him to St. Louis, but Congress had not yet acted on the annual appropriations. On May 4, after having waited for two weeks in St. Louis, Long and his companions set out for Council Bluff on horseback.

> EJ: As we followed the little pathway [to the west], we passed large tracts to which the labours of the sand rat had given the aspect of a ploughed field. From the great quantities of fresh earth recently brought up, we perceived the little animals were engaged in enlarging their subterranean excavations; and we watched long, though in vain, expecting to see them emerge from their burrows.

From these remarks, it is unclear whether James collected specimens at this locality, but a long footnote in the *Account*, doubtless prepared by Thomas Say, suggests that specimens were taken somewhere on the Missouri. Say identified the animal as *Mus bursarius*, the plains pocket gopher. He described the unique fur-lined cheek pouches as well as other anatomical details, and proposed that the animal be placed in a previously undescribed genus, which he named *Pseudostoma* (Greek for "false mouth"). Here, as elsewhere, he seemed unfamiliar with the publications of the eccentric naturalist Constantine Rafinesque, who had described the genus in 1817, calling it *Geomys* (Greek for "earth mouse"). Although both generic names are reasonably descriptive, Rafinesque's, of course, has priority and is the name now used.

> EJ: The whole animal has a clumsy aspect, having a large head and body, with short legs, large fore feet, and small hind feet; and although it walks awkwardly, yet it burrows with the greatest rapidity, so that the difficulty of obtaining specimens may be, in a great degree, attributed to the facility with which the animal passes through the soil, in removing from the vicinity of danger.

The next day the three men passed a pond in which "the nelumbium was growing in great perfection," its blossoms larger than those of any other North American plant except the magnolia. This was American lotus, also called pond nut, water chinquapin, or wonkapin (*Nelumbo lutea*). The nuts, James reported, "have, when ripe, the size and the general appearance of small acorns, but are much more palatable. The large farinaceous root is sometimes used by the Indians as an article of diet, as are also the nuts."

Farther on, James noted several other interesting plants, including spiderwort (*Tradescantia*), alumroot (*Heuchera*), ninebark (*Physocarpus*), and false mermaid (*Floerkea*). The forests were composed of maple, beech, ash, basswood, and hop hornbeam (*Ostrya virginiana*), all attaining "an uncommon magnitude."

On May 8, the men arrived at Franklin, where they waited in vain for another five days for funds and further instructions from

Washington. These delays would mean that the trip west could not begin before early June, rather than early May, as originally planned. The plains would be scorchingly hot, and the work planned for four months would have to be squeezed into three.

Having decided that he could wait no longer, Long and his companions set out from Franklin cross-country, by horseback. It was an uncomfortable trip. They frequently camped in the rain, without shelters other than those they could improvise on the spot. Deer and elk were plentiful, but the men had little success in taking any. They often had to make do with wild onions, lamb's-quarter pigweed, and roast squirrel. At night their sleep was interrupted by "the hooting of owls, together with the howling of wolves, and the cries of other nocturnal animals." A collection of plants that James had made was destroyed in a storm. Mosquitoes and ticks were relentless in their attacks. But there were moments of pleasure, as when they encountered "two plants of singular beauty": great flowering penstemon (*Penstemon grandiflora*) and Colorado locoweed (*Oxytropis lambertii*). There were barn swallows, turkeys, prairie-chickens, and sandhill cranes; Bell found that the cranes made "a very disagreeable noise."

Eventually they arrived at Engineer Cantonment, to be greeted by Say, Graham, and Seymour, "dressed in leather hunting shirts and leggens . . . as if they had assumed the dress and appearance of the hunter from choice or singularity," wrote Bell. The next day Long and Bell rode to Camp Missouri to arrange for men and supplies for their trip west, though scurvy and privation had taken their toll at the camp, and they received fewer men and provisions than they had hoped. On June 1, Long issued orders for the following several months.

SL: Agreeably to the instructions of the Honorable Secretary of War, the further progress of the Exploring Expedition up the Missouri is arrested during the present season. By the same authority, an excursion, by land, to the sources of the

Titian Peale, pencil sketch of a prairie-chicken. (American Philosophical Society)

river Platte, and thence by way of the Arkansa and Red rivers to the Mississippi, is ordered. The expedition will accordingly proceed on this duty as soon as practicable. . . . The duties [formerly] assigned to Major Biddle will be performed by Captain J. R. Bell . . . with the exception of those parts which relate to the manners, customs, and traditions of the various savage tribes which we may pass. The duties thus excepted will be performed by Mr. Say. The duties assigned to Dr. Baldwin and Mr. Jessup . . . will be performed by Dr. E. James. . . . In these duties are excepted those parts which relate to comparative anatomy, and the diseases, remedies, etc. known amongst the Indians; which will also be performed by Mr. Say.

Lieutenant Graham will take charge of the United States' steam-boat, Western Engineer, and proceed down the Missouri to the Mississippi, with the remaining part of the crew. . . .

The detachment from the rifle regiment . . . will accompany the expedition . . . under the immediate command of Lieutenant Swift. . . .

The duties of the expedition being arduous, and the objects in view difficult of attainment, the hardships and exposures to be encountered requiring zealous and obstinate perseverance, it is confidently expected, that all embarked in the enterprise will contribute every aid in their power, tending to a successful and speedy termination of the contemplated tour.

The party was now quite different from the one that had set out from Pittsburgh thirteen months earlier. Of the twenty-two men, only five were holdovers from the previous year.

MILITARY
Major Stephen H. Long	Commander
Captain John R. Bell	Journalist
Lieutenant William H. Swift	Assistant topographer
Corporal William Parish	
Private John Sweney	
Private Joseph Verplank	
Private Robert Foster	
Private Mordecai Nowland	
Private Peter Barnard	
Private Charles Myers	

SCIENTIFIC
Dr. Edwin James	Botanist, geologist, and physician
Thomas Say	Zoologist and ethnologist
Titian R. Peale	Assistant naturalist
Samuel Seymour	Artist

SUPPORT STAFF
H. Dougherty (brother of John Dougherty)	Hunter
Zachariah Wilson	Baggage master
Stephen Julien	French and Indian interpreter
D. Adams	Spanish interpreter
James Oakley	Engagee
James Duncan	Engagee

Two others joined the expedition at the Pawnee villages, in Nebraska, where they lived.

Abraham Ledoux	Hunter and interpreter
Joseph Bijeau	Guide and interpreter

Bell's journal includes Long's muster roll, in which the salaries of the expedition members were listed. James and Say received the highest pay, $2.20 a day; Peale and Seymour got $1.70 a day; Long, Bell, and Swift earned $1.50 a day. The privates received 15 cents a day. Long was to have received $2,000 from the War Department to purchase supplies, but this money still had not arrived when the time came to leave. Consequently, the expedition was remarkably poorly equipped, and the naturalists had in some cases to acquire their own horses and necessities. The supplies they took for trading with the Indians fell especially short of what was needed.

> EJ: Our outfit comprised the following articles of provisions, Indian goods, &c. viz. 150 lb. of pork, 500 lb. of biscuit, 3 bushels of parched corn meal, 5 gallons of whiskey, 25 lb. of coffee, 30 lb. of sugar, and a small quantity of salt, 5 lb. of vermilion, 2 lb. of beads, 2 gross of knives, 1 gross of combs, 1 dozen of fire steels, 300 flints, 1 dozen of gun worms, 2 gross of hawk's bells, 2 dozen of mockasin awls, 1 dozen of scissors, 6 dozen of looking glasses, 30 lb. of tobacco, and a few trinkets, 2 axes, several hatchets, forage-bags, canteens, bullet-pouches, powder-horns, tin cannisters, skin canoes, packing-skins, pack cords, and some small packing-boxes for insects, &c.

In addition, there were compasses, sextants, and other "instruments for topographical purposes." Soldiers and hunters carried rifles. Ammunition included "about 30 pounds of powder, 20 pounds of balls, and 40 pounds of lead, with a plentiful supply of flints, and some small shot."

There were twenty-eight horses and mules, one for each man to ride and several for carrying packs. Crude saddles were made by the men or purchased from Indians. Two dogs accompanied the expedition: Caesar and Buck. Both became favorites of the men, as dogs will do.

Before their departure, members of the expedition were treated to a banquet at Camp Missouri.

> JUNE 2. JB: [A]t 2 p.m. we set down to an excellent dinner—consisting of almost all the varieties of tame and wild meats and fowls—and garden vegetables except potatoes—pastry pies, of dried apple and gooseberries—of liquors we had Madeira wine, brandy, rum & whisky—no person could hardly suppose a table could be furnished with such a variety in the wilderness 600 miles by land west of the Mississippi river.... [T]he greatest order and harmony prevailed—after the cloth was removed we had some good songs, patriotic tosts & sentiments from the officers, accompanied by the music of the excellent band belonging to the 6th Regiment. We returned in the evening highly gratified with business of the day.

The explorers had good reason to wonder what lay ahead of them as they went about preparing themselves to traverse country that had never been adequately mapped and was inhabited only by Indians.

> EJ: Several of the Indians about Council Bluff, to whom our proposed route had been explained, and who had witnessed our preparations, affected to laugh at our temerity, in attempting what they said we should never be able to accomplish. They represented some part of the country, through which we intended to travel, as so entirely destitute of water and grass, that

neither ourselves nor our horses could be subsisted while passing it. Baron Vasquez, who accompanied Captain Pike, in his expedition to the sources of the Arkansa, assured us there was no probability we could avoid the attacks of hostile Indians, who infested every part of the country. The assault which had been recently made by a party of the Sauks and Foxes, upon a trading post . . . above Council Bluff, in which one man was killed, and several wounded, had at this time spread considerable terror among those in any degree exposed to the hostilities of the Indians.

With these prospects, and with the very inadequate outfit above described, which was the utmost our united means enabled us to furnish, we departed from Engineer Cantonment, at 11 o'clock, on the 6th of June.

They would be gone until early September—only three months to cover more than 1,500 miles by foot and horseback. Clearly the naturalists would be hard put to gather data and specimens at a pace averaging about 15 miles a day, but requiring days of 20 miles or more to compensate for Sundays (when Long preferred not to travel) and for unforeseen delays.

James's *Account* is now based in part on his personal observations, rather than being only a compilation from journals of other expedition members. Bell's somewhat less literate *Journal* provides a good day-to-day account of the party's movements; he clearly appreciated the efforts of the naturalists, and was occasionally eloquent on beauties and possibilities of the landscapes through which they passed.

Engineer Cantonment would become a thing of the past, but Camp Missouri continued to be garrisoned for several years, though its name was changed to Fort Atkinson. In 1825, Henry Atkinson (now a general) made a second attempt to reach the mouth of the Yellowstone. This time he succeeded, using eight keelboats powered by poles, sails, ropes hauled from the shore, and paddle wheels operated by muscle power. Major Benjamin O'Fallon accompanied him, and treaties were signed with several Indian tribes. The expedition returned from the Yellowstone in the fall of the same year, without having established a post there. Atkinson was able to report that "not a boat or man was lost"—in contrast to the 1819/1820 expedition, when so much time was consumed in going half the distance, leaving his steamboats behind and watching many of his men succumb to scurvy. Fort Union was finally established at the mouth of the Yellowstone in 1828, but by the American Fur Company rather than the military.

But we must follow Stephen Long and his men as they headed west across the prairies. It is especially regrettable that they were unable to leave until June 6, when the flush of spring was already past. They would have to move quickly to be back before the first snows began to fall, giving the naturalists much less time than they would have preferred to collect and prepare specimens and to write up their notes.

Fortunately Say and his colleagues were able to send many specimens, as well as their field notes from 1819 and the early spring of 1820, down the Missouri with the *Western Engineer*, and most of them reached Philadelphia safely. They were to be less fortunate after their trek to the Rockies in 1820.

$\mathcal{S}\!even$
TO THE ROCKIES

ON THE FIRST DAY OUT FROM Engineer Cantonment, the expedition moved only a few miles, camping for the night on Papillion Creek. There were delays occasioned by "the derangement of the packs, the obstinacy of the mules, and the want of dexterity and experience in our engagees." It rained in the night, but three tents adequately protected the men. On the following morning they reached the Elkhorn, a river about thirty yards wide.

JUNE 7. EJ: At this time our horses were barely able to keep their feet, in crossing the deepest part of the channel. Our heavy baggage was ferried across in a portable canoe, consisting of a single bison hide, which we carried constantly with us. Its construction is extremely simple; the margin of the hide being pierced with several small holes, admits a cord, by which it is drawn into the form of a shallow basin. This is placed upon the water, and is kept sufficiently distended by the baggage it receives; it is then towed or pushed across. A canoe of this kind will carry from four to five hundred pounds. . . .

A species of onion, with a root about as large as an ounce ball, and bearing a conspicuous umbel of purple flowers, is very abundant about the streams, and furnished a valuable addition to our bill of fare.

Soon after crossing the Elk-horn we entered the valley of the Platte, which presented the view of an unvaried plain, from three to eight miles in width, and extending more than one hundred miles along the river, being a vast expanse of prairie, or natural meadow, without a hill or other inequality of surface, and with scarce a tree or a shrub to be seen upon it. The wood-lands, occupying the islands in the Platte, bound it on one side; the river-hills, low and gently sloped, terminate it on the other.

During travel, Bell and one of the guides rode first, followed by the soldiers and support staff in a single file, with the packhorses. Long brought up the rear. The "scientific gentlemen" were allowed to occupy "any part of the line that suited their convenience." On the evening of June 7, one of the men caught a young pronghorn antelope, which Peale sketched and then released. Bell recorded that they had covered sixteen miles that day.

On the following day, there was a violent storm, during which lightning struck the ground a short distance away, throwing "water and mud . . . several feet into the air by the shock." After covering twenty-four miles, the explorers camped on Coquille Creek, where mosquitoes swarmed "in inconceivable multitudes."

The next day brought them to the "Loup fork of the Platte" (Loup River), where James noted several plants that he had not seen previously. One was a mallow "with a large tuberous root which is soft and edible, being by no means ungrateful to the taste." Others included a plantain (*Plantago*), a puccoon (*Lithospermum*), a milk-vetch (*Astragalus*), a true vetch (*Vicia*), and "the superb sweet pea" (hoary vetchling, *Lathyrus polymorphus*). Also noted were prairie false dandelion, rough pennyroyal, scarlet gaura, and several grasses. The naturalists also saw curlews, marbled godwits, and upland sandpipers.

On June 10, both Say and Seymour were thrown from their

horses while crossing streams, and Say lost some of his equipment. The next day brought them to the first of three villages of the Pawnee Indians, which they visited in succession. Long and his men camped nearby, where at night the dogs of the Indians "howled in concert, in the same voice, and nearly the same tune, as the wolves, to whose nightly serenade we were now accustomed." The Pawnees offered them dried corn, pumpkins, and jerked bison meat in exchange for tobacco, vermilion, beads, and mirrors. Once again, the party was warned that their trip "was attended with great difficulties and danger" because of the lack of water and game as well as the presence of "bands of powerful and ferocious Indians." The Pawnees' design, wrote James, was "to deter us from passing through their hunting grounds, and perhaps hoping by these means to possess themselves of a larger share of the articles we had provided for Indian presents."

> EJ: The three Pawnee villages, with their pasture grounds and insignificant enclosures, occupy about ten miles in length of the fertile valley of the Wolf [Loup] river. The surface is wholly naked of timber, rising gradually to the river hills, which are broad and low, and from a mile to a mile and a half distant. The soil of this valley is deep and of inexhaustible fertility. The surface, to the depth of two or three feet, is a dark coloured vegetable mould intermixed with argillaceous loam, and still deeper, with a fine siliceous sand. The agriculture of the Pawnees is extremely rude. They are supplied with a few hoes by the traders, but many of their labours are accomplished with the rude instruments of wood and bone which their own ingenuity supplies. They plant corn and pumpkins in little patches along the sides of deep ravines, and wherever by any accident the grassy turf has been eradicated.

In 1806, Pike had estimated that the three villages together had a population of 6,223; Bell estimated that there were now about 8,000 inhabitants. The tribes had 6,000 to 8,000 horses. In winter, the

Pawnees left their villages and moved to wooded valleys, not only to find fuel but also to feed their horses, which survived well on the small branches and inner bark of cottonwoods.

James noted that the Indians made much use of the roots of *Psoralea esculenta*, a legume often called scurfpea or breadroot, and known to Canadian trappers as *pomme blanche*. "It is eaten either boiled or roasted," wrote James, "and somewhat resembles the sweet potatoe." This was a popular food among the Indians of the Missouri and Platte Valleys. John Colter is said to have subsisted on *pomme blanche* for seven days during his flight from the Blackfeet Indians in 1810, before reaching safety in one of Manuel Lisa's posts.

Like many tribes, the Pawnees had suffered from smallpox, and Long had been asked to introduce the concept of vaccination to them. Vaccine had been sent to St. Louis by mail, and then carried up the Missouri by keelboat and on to the Pawnee villages by messenger. The Pawnees were understandably reluctant to be vaccinated, even though Long and others allowed themselves to be vaccinated in the presence of the Indians. The whole incident seems ludicrous, since the vaccine had been soaked in water when the keelboat carrying it was wrecked, and the vaccine was known to be "unfit for use."

The explorers once again met with the chief, Latelesha, and his son Petalesharoo. Bell described Latelesha as "a fine looking man, large & fat—with a good deal of goodness & friendship expressed in his countenance." At the chief's lodge, the men were fed "bowls of corn & buffalo guts boiled." "[T]his dish relished well & we eat heartily of it," reported Bell. The next day, Latelesha attempted to obtain whiskey from Long, but was refused, whereupon he left "without taking leave or shaking hands."

The men were visited by a medicine man who was curious to know how the whites "made medicine." He was shown a case of surgeon's instruments, whose uses were explained, but he "at length turned abruptly away, with an air of dissatisfaction and contempt."

At the Pawnee villages, Say collected "a large and beautiful

Titian Peale, drawing of a bison-skin robe depicting a battle between the Pawnees (*right*) and the Kansas Indians (*left*). The robe was presented to the expedition by the Pawnees, and the illustration appeared in the *Account*.

insect" that he later described as *Melolontha* (now *Polyphylla*) *10-lineata*. This striking scarab beetle, about an inch long and with huge antennae and a pattern of white stripes down its back, is now known to be widely distributed in the West. Its larvae are "white grubs" that feed on the roots of shrubs and trees.

By prior arrangement, two men were to join the expedition at the Pawnee villages, to serve as guides, interpreters, and hunters. But the men were reluctant to leave their homes and families. When Long threatened that the Indian agent might deny them the opportunity to live and trade with the Pawnees in the future, the Canadians Joseph Bijeau and Abraham Ledoux consented to join. Both were to prove indispensable.

On June 13, Long and his men crossed the Loup River, but with some difficulty, as the current was strong and the bottom "par-

Striped June beetle, first taken by Thomas Say at the Pawnee villages.

took something of the nature of quick-sands." Long, Say, and others were thrown from their horses; much of their equipment fell into the water, and some was lost. This was Say's second immersion, and he was now "in great measure, unencumbered with baggage." Along the bank, James noted "a large flowering rose . . . diffusing a most grateful fragrance." Another attractive shrub James suspected to be *Symphoria glomerata* (it was later described as *Symphoricarpos occidentalis*, wolfberry or western snowberry). On the hills grew prickly pear cacti (*Cactus fragilis*, now *Opuntia fragilis*).

As the men left the Pawnee villages, they passed colonies of prairie dogs, or "Louisiana marmots." "This interesting and sprightly little animal," wrote James, "has received the absurd and inappropriate name of Prairie dog, from a fancied resemblance of its warning cry to the hurried barking of a small dog." It was apparently Meriwether Lewis who first likened their calls to those of "little toy dogs," and his companion Sergeant John Ordway who first consistently called them "prairie dogs," though Lewis preferred "barking squirrel" and Clark "burrowing squirrel," both more appropriate names.

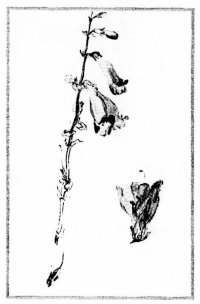

Titian Peale, sketch of a flower, evidently a beardtongue, near the Pawnee villages, June 12, 1820. (From the sketchbooks of Titian Ramsay Peale, Yale University Art Gallery, gift of Ramsay MacMullen, M.A.H. 1967)

Titian Peale, sketch of a flower, perhaps a puccoon, near the Pawnee villages, June 12, 1820. (From the sketchbooks of Titian Ramsay Peale, Yale University Art Gallery, gift of Ramsay MacMullen, M.A.H. 1967)

James's *Account* includes a description of the "dogs'" anatomy and behavior, doubtless provided by Say.

> EJ (PROBABLY AFTER TS): They delight to sport about the entrance of their burrows in pleasant weather; at the approach of danger they retreat to their dens; or when its proximity is not too immediate, they remain, barking, and flourishing their tails, on the edge of their holes, or sitting erect to reconnoitre. When fired upon in this situation, they never fail to escape, or if killed, instantly to fall into their burrows, where they are beyond the reach of the hunter.
>
> The burrows are not always equidistant from each other, though they occur usually at intervals of about twenty feet.

A day or two later, two prairie dogs were taken, roasted, and found to be "well flavoured," rather like woodchuck. Lewis and Clark deserve credit for the discovery of the black-tailed prairie dog. They sent a live individual, accompanied by four live magpies and a sharp-tailed grouse, from Fort Mandan (in present-day North Dakota) to Thomas Jefferson in 1805, along with the skins and skulls of other animals. The shipment left on a keelboat to St. Louis, and then traveled by barge to New Orleans. The live animals had been fed along the way, but even so the grouse had died and the prairie dog was in poor condition on arrival in New Orleans after a trip of 2,500 miles from Fort Mandan. With care, the prairie dog improved and with the magpies was sent to Baltimore on the schooner *Comet*. Three of the four magpies died at sea, but the prairie dog and the remaining magpie reached Jefferson, who studied them and sent them on to the Philadelphia Museum (formerly Peale's Museum, and now housed in Independence Hall). As Paul Russell Cutright has remarked, never before or since have live, wild animals resided in both the president's mansion and Independence Hall. The prairie dog survived for a few more months, but George Ord's formal description of the animal, in 1815, was drawn from skins and skeletons.

Titian Peale, sketches of horned lizards ("orbicular lizards") and prairie dogs at their "village." Both animals were encountered several times at diverse localities. (From the sketchbooks of Titian Ramsay Peale, Yale University Art Gallery, gift of Ramsay MacMullen, M.A.H. 1967)

Ord chose the name *Cynomys ludovicianus* (literally, dog-mouse of Louisiana Territory) for these animals, so characteristic of major parts of the West.

James noted that the soil around prairie dogs' burrows was the particular habitat of a species of *Solanum* (buffalo bur). He also described a composite from this site, calling it *Hieracium runcinatum* (now *Crepis runcinata*, hawksbeard).

JUNE 14. EJ: On arriving near the Platte we observed a species of prickly pear (*Cactus ferox*. N.) [now *Opuntia polyacantha*] to become very numerous. . . . Our Indian horses were so well acquainted with this plant, and its properties, that they used the utmost care to avoid stepping on it. The flowers are of a sulphur yellow, and when fully expanded are nearly as large as those of the garden paeony. . . . A second species, the *C. mamillaris*, N. [now *Coryphantha missouriensis*] occurs on the dry sandy ridges between the Pawnee villages and the Platte. [Here, as in many places, James followed the names in Thomas Nuttall's *Catalogue*, hence the "N" after the species names.] The beautiful cristaria *coccinea* [now *Callirhoe involucrata*, purple poppy mallow] is very frequent in the low plains along the Platte. Its flowers have nearly the aspect of those of the common wild rose, except that they are more deeply coloured.

The expedition soon arrived at Grand Island, then uninhabited country, though now the site of Nebraska's third largest city. The hunters were unsuccessful in taking bison or pronghorn antelope. On June 16, according to Bell, the party was "put on an allowance of one biscuit" per day, to which was added a small quantity of bologna, brought by the *Western Engineer* from Pittsburgh the previous year and "too highly seasoned to eat off hand." But in the evening, the hunters took a pronghorn antelope, which was welcome though not generous when divided twenty-two ways.

> EJ: The antelope possesses an unconquerable inquisitiveness, of which the hunters often take advantage, to compass the destruction of the animal. The attempt to approach immediately towards them in the open plain, where they are always found, rarely proves successful. Instead of this, the hunter, getting as near the animal as is practicable, without exciting alarm, conceals himself by lying down, then fixing a handkerchief or cap upon the end of his ramrod, continues to wave it, still remaining concealed. The animal, after a long contest between curiosity and fear, at length approaches near enough to become a sacrifice to the former.

The pronghorn seems fated to be called an antelope, even though it bears no close relationships to the antelopes of the Old World. In fact, it is the unique representative of a group of ungulates that, in prehistoric times, was represented by several species that roamed North America. In the early nineteenth century, there may have been more than 40 million of these graceful animals throughout the West. By 1900, there were few remaining, but thanks to conservation efforts they may now be seen along many western highways. It was

Say's friend George Ord who named the species *Antelope* (now *Antilocapra*) *americana*, on the basis of specimens collected by Lewis and Clark.

James meanwhile was botanizing further. *Cherianthus asper* (now *Erysimum asperum*, western wallflower), he found, "is intensely bitter in every part, particularly the root, which is used as medicine by the Indians." Other plants noted were frostwort, stickseed, slender beardtongue, and sideoats grama grass. In ponds along the Platte, James found pondweed (*Potamogeton*), bladderwort (*Utricularia*), and water milfoil (*Myriophillum*).

Almost every day there was a storm, and sometimes the wind was so strong that only by standing outside the tents and holding them to the ground could the men prevent the tents from blowing over. Anyone who has camped on the plains of Nebraska can appreciate their problems!

Sunday, June 18, was devoted to washing clothes, organizing packs, and cutting up and drying antelope meat. The horses were permitted to graze. James studied his specimens, but found that the papers he had carried for pressing plants had been insufficiently protected from the weather, "some of our collections being in part wet, and others having been made during heavy rains." The area around the camp was strewn with the bones of bison and other animals, and there were a few human skulls, suggesting that a massacre had taken place there. One of the skulls "we thought it no sacrilege to compliment with a place upon one of our packhorses."

The explorers were about 200 miles up the Platte River. "It was still from one to three miles in breadth, containing numerous islands, covered with a scanty growth of cotton wood [and] willows, the amorpha fruticosa [false indigo], and other shrubs." Another common plant was wild licorice (*Glycyrrhiza lepidota*); the root of which in taste "bears a very slight resemblance to the liquorice of the shops, but is bitter and nauseous."

On the next day, the expedition moved another thirty miles along the north shore of the Platte, and James reported prickly pop-

Titian Peale, sketch of a plant, evidently a heliotrope (perhaps *Heliotropium curassavicum*), June 21, 1820. It was found by Edwin James in saline soil, near the confluence of the North and South Forks of the Platte. (From the sketchbooks of Titian Ramsay Peale, Yale University Art Gallery, gift of Ramsay MacMullen, M.A.H. 1967).

pies (*Argemone*), with broad white flowers and stems that when broken exuded a bitter, yellow fluid. There were yuccas, "thriving with an appearance of luxuriance and verdure, in a soil which bids defiance to almost every other species of vegetation." Other plants noted were several kinds of milkweeds (*Asclepias*), sunflowers (*Helianthus*), and species of daisies (*Erigeron*), dock (*Rumex*), speedwell (*Veronica*), and skullcap (*Scutellaria*).

In drying pools of water there were tadpole shrimps, "great numbers of which were dying upon the surrounding mud." The naturalists noted that the shrimps' dorsal shields made them resemble miniature horseshoe crabs. There were "about sixty pairs of feet, and [they] swim upon their back," noted Say. These crustaceans are known to produce drought-resistant eggs that hatch when the temporary pools are once again filled with water. They are very ancient animals, apparently differing hardly at all from fossils of 170 million years ago. Long's naturalists may have been the first to have reported them from North America. Say named the shrimps *Apus obtusus*, but his description was too brief to properly characterize the species and he brought back no specimens. It is now called *Triops longicaudus* (literally, three eyes with a long tail). These crustaceans do not really have three eyes, but they do have a "dorsal organ" resembling a third eye, another feature (along with the many legs and the horseshoe-shaped shield) suggesting that indeed these are animals surviving from the distant past.

JUNE 20. JB: Allowance to two thirds of a biscuit per day—supper last night, & breakfast this morning without meat. This determined the commanding officer to remain encamped during the day [although it was Tuesday]—and that every exertion should be made by the hunters & marksmen of the party to take some kind of game—accordingly every man, that had any pretensions to being a good shot went out—and returned by the middle of the afternoon, having killed one buffalo, three antelopes, and one hare—after that all hands was engaged in jerking and drying the meat, to take it with us. To jerk meat, is to cut it in thin flakes; to dry it, is to place on a frame consisting of small poles under which is kept a moderate fire—in this way the meat is cured without salt.

The men were now in country where firewood was not always readily available, and they often used dried bison dung as fuel. On June 22, they arrived at the confluence of the north and south forks of the Platte. There they watched two elk crossing the north fork, and they made their own crossing safely at the same point.

It was Long's intention to follow the South Platte to the mountains. Bijeau, the guide, had followed both forks of the Platte, and he reported that the sources of the North Platte were small streams arising in "a circumscribed valley within the mountains, called the Bull-pen. This basin is surrounded by high and rugged mountains, except at the place where the north fork passes into the plains." The width of the valley was said to be about twenty miles. This is the first description in print of what is now called North Park, a broad and glorious valley rimmed by the Park and Medicine Bow ranges. It now makes up the better part of Jackson County, Colorado.

> EJ: Three beavers were seen cutting down a large cotton-wood tree; when they had made considerable progress, one of them retired to a short distance, and took his station in the water, looking steadfastly at the top of the tree. As soon as he perceived the top begin to move towards its fall, he gave notice of the danger to his companions, who were still at work, gnawing at its base, by slapping his tail upon the surface of the water, and they immediately ran from the tree out of harm's way.

The naturalists speculated as to whether the American beaver belonged to the same species as its European counterpart (it is now considered distinct). This is one of the few discussions in James's *Account* of the beaver, an animal that had played, and was still to play, a major role in the opening of the West to people of European descent.

Thus far, the men had seen only an occasional bison, but this was about to change.

> JUNE 23. EJ: Our view of the opposite margin of the Platte, during this day's march, had been intercepted by an elevated swell of the surface, which extended along, parallel to the river, that we were now approaching. Immediately upon surmounting this undulation we saw before us, upon the broad expanse of the left margin of the river, immense herds of bison, grazing in undisturbed possession, and obscuring, with the density of their numbers, the verdant plain; to the right and left, as far as the eye was permitted to rove, the crowd seemed hardly to diminish, and it would be no exaggeration to say, that at least ten thousand here burst on our sight in the instant. Small columns of dust were occasionally wafted by the wind from the bulls that were pawing

the earth, and rolling; the interest of action was also communicated to the scene, by the unwieldy playfulness of some individuals, that the eye would occasionally rest upon, their real or affected combats, or by the slow or rapid progress of others to and from their watering places. On the distant bluffs, individuals were constantly disappearing, while others were presenting themselves to our view, until, as the dusk of the evening increased, their massive forms, thus elevated above the line of other objects, were but dimly defined on the skies. We retired to our evening fare, highly gratified with the novel spectacle we had witnessed, and with the most sanguine expectations of the future.

In the morning we again sought the living picture, but upon all the plain which last evening was so teeming with noble animals, not one remained. We forded the [South] Platte with less delay and difficulty than we had encountered in crossing the north fork. . . .

We had no sooner crossed the Platte, than our attention was arrested by the beautiful white primrose (oenothera pinnatifida, N.) with its long and slender corolla reclining upon the grass. The flower, which is near to two inches long, constitutes about one-half of the entire length of the plant.

This evening primrose is now called *Oenothera coronopifolia*. Other plants collected near the site included white and purple prairie clover (*Dalea candida* and *D. purpurea*); species of spider flower, or bee plant (*Cleome*); beardtongue (*Penstemon*); and milkvetch (*Astragalus*).

Seymour and Peale sketched bison, now so plentiful. The men preferred bison meat to that of elk or deer, "which was thrown away when it could be substituted by the bison meat." They also saw hares, badgers, coyotes, eagles, vultures, ravens, and owls. These, wrote James, "in some measure relieved the uniformity of [the plain's] cheerless scenery."

Titian Peale, watercolor of a bison. This is believed to be the earliest illustration of a bison grazing on the plains. (American Philosophical Society)

EJ: Some extensive tracts of land along the Platte . . . are almost exclusively occupied by a scattered growth of several species of wormwood. . . . The peculiar aromatic scent, and the flavour of this well known plant, is recognized in all the species. . . . Several of them are eaten by the bisons, and our horses were sometimes reduced to the necessity of feeding upon them.

Wormwood was undoubtedly well known to James from his medical training, since various kinds have long been used to remove worms from the intestinal tract. More often called sagebrush, these plants cover vast acreages of the West. The dominant species along the South Platte was very likely silvery wormwood, or sandhill sage (*Artemisia filifolia*), a species described by John Torrey in 1828 from specimens collected by James.

Over the next few days, the weather became cool and pleasant, and the expedition members began to anticipate their first view of the mountains. Early on the morning of June 26, a gun was discharged, a signal that Indians were approaching. But Long was merely testing his men's readiness. James confessed that although the men were now used to rising well before sunrise, they "still found we left that small spot of earth, on which we had rested our limbs, and which had become warm and dry by the heat of our bodies, with as much reluctance as we have felt at quitting softer beds."

They were moving rapidly up the south side of the South Platte, covering, according to Bell, twenty-four miles on June 26, forty miles on June 27, and twenty-seven miles on June 28. They had now crossed into what is now Colorado, near the present city of Julesburg. Bison continued to be abundant, and James was led to muse on their ultimate fate.

> EJ: It would be highly desirable that some law for the preservation of game might be extended to, and rigidly enforced in the country where the bison is still met with; that the wanton destruction of these valuable animals, by the white hunters, might be checked or prevented. It is common for hunters to attack large herds of these animals, and having slaughtered as many as they are able, from mere wantonness and love of this barbarous sport, to leave the carcasses to be devoured by the wolves and birds of prey; thousands are slaughtered yearly, of which no part is saved except the tongues. This inconsiderate and cruel practice is undoubtedly the principal reason why the bison flies so far and so soon from the neighbourhood of our frontier settlements.

The naturalists of the Long Expedition were among the first to decry the slaughter of the bison. Twenty years later, Audubon hunted bison on the upper Missouri. He noted that the herds were already declining, and remarked that the species might disappear "like the Great Auk" unless something was done to prevent it. But a few years later, "Buffalo

Bill" Cody bragged that he had killed 4,280 bison in one year, while he worked for the Union Pacific Railroad. The railroad crossed land owned by the Sioux, and it was easier to destroy their way of life than to defeat warriors such as Sitting Bull and Crazy Horse. "No sight is more common on the plains," wrote Theodore Roosevelt in 1888, "than that of a bleached buffalo skull." Now even the skulls are gone, and the few remaining herds are semidomesticated.

Meanwhile, as the expedition trekked through what is now northeastern Colorado, zoologist Thomas Say was turning his attention to small game: beetles. One of these was a blister beetle, which he named *Lytta albida* (now *Macrobasis albida*); another, a darkling beetle (*Blaps obscura*, now *Eleodes obscura*). Both were illustrated in color in the first volume of his *American Entomology*, based on drawings by Titian Peale. The blister beetle that Say described as a "remarkably fine species" was taken as the expedition passed over "that vast desert." "It appeared to be feeding upon the scanty grass, in a situation from which the eye could not rest upon a tree, or even a humble shrub, throughout the entire range of vision, to interrupt the uniformity of a far outspreading, gently undulated surface, that, like the ocean, presented an equal horizon in every direction."

> JUNE 27. EJ (DOUBTLESS FROM TS): We observed, in repeated instances, several individuals of a singular genus of reptiles . . . which in form resemble short serpents, but are more closely allied to the lizards, by being furnished with two feet. They were so active, that it was not without some difficulty that we succeeded in obtaining a specimen. Of this (as was our uniform custom, when any apparently new animal was presented) we immediately drew out a description. But as the specimen was unfortunately lost, and the description formed part of the zoological notes and observations, which were carried off by our deserters, we are reduced to the necessity of merely indicating the probability of the existence of the [two-legged lizard] within the territory of the United States.

These observations have been the cause of discussion among herpetologists ever since. Two-legged lizards (now *Bipes*, in the family Amphisbaenidae) have never been found in Colorado again, and only doubtfully recorded from north of Mexico. They are slow-moving, wormlike, mostly subterranean animals. But the members of the expedition found them so active that they had trouble catching them. Perhaps what they actually saw were many-lined skinks (*Eumeces multivirgatus*), slender, fast-running lizards that have four unusually short legs. In his book *Amphibians and Reptiles of Colorado*, Geoffrey Hammerson states, "I cannot seriously consider *Bipes* to be an inhabitant of Colorado until firm documentation of its occurrence is obtained."

The comments concerning the discovery of the two-legged lizard are doubly interesting because they explain the naturalists' practice of drawing up a preliminary description in the field. In this case, the specimen was lost and the description disappeared when many of Say's notes and collections were carried off by deserters on the return trip along the Arkansas River. Under the circumstances, Say was unable to include a description in the *Account* of the expedition.

On June 28, the party saw and admired a herd of wild horses. "Their playfulness," wrote James, "rather than their fears, seemed to be excited by our appearance, and we often saw them, more than a mile distant, leaping and curvetting, involved by a cloud of dust, which they seemed to delight in raising."

They also saw several small foxes and shot some for specimens. Say described them in detail and provided a name, *Canis velox* (now *Vulpes velox*). *Velox* is the Latin word for "swift," and this beautiful animal is now called the swift fox. This fox, wrote Say, "runs with extraordinary swiftness, so much so, that when at full speed its course has been by the hunters compared to the flight of a bird skimming the surface of the earth. . . . [I]t burrows in the earth, in a country totally destitute of trees or bushes, and is not known to dwell in forest districts." Once again, the description that Say drew up was stolen by the deserters. It was later reconstructed in part from a head that the naturalists had managed to preserve.

The swift fox is now gone from much of its former range, as it is easily trapped and is often the victim of control efforts directed toward coyotes. The foxes feed primarily on insects and small rodents and are therefore beneficial to humans. But since they are classed as predators, they are not tolerated by ranchers, whose cattle have in any case degraded much of the swift fox's habitat.

The expedition now passed "Cherry Creek," which "heads in the Rocky Mountains." This was, of course, not modern Cherry Creek, which flows through Denver. It may have been modern Paw-nee Creek or more likely (according to George Goodman and Cheryl Lawson) Cedar Creek. Both arise on the High Plains well east of the Rockies. Magpies were seen on islands in the river. These birds, so characteristic of the West, actually range through Alaska and on through much of Eurasia. Lewis and Clark had seen them on the upper Missouri, calling the bird "a butifull thing," as indeed it is.

> JUNE 29. EJ: The country . . . is as uniformly plain as that on any part of the Platte. It differs from that further to the east only in being of a coarser sand, and in aspect of more unvaried sterility. The cactus ferox [*Opuntia polyacantha*] reigns sole monarch, the sole possessor, of thousands of acres of this dreary plain. It forms patches which neither horse nor any other an-imal will attempt to pass over. . . . In depressed and moist sit-uations, where the soil is not so entirely unproductive, the variegated spurge (euphorbia variegata) [*Euphorbia marginata,* snow-on-the-mountain], with its painted involucrum and parti-coloured leaves, is a conspicuous and beautiful ornament. The lepidium virginicum [pepper-grass] . . . is here of such diminu-tive size that we were induced to search, though we sought in vain, for some character to distinguish it as a separate species.

JUNE 30. JB: [A]t 8 o'clock, being on an elevated part of the prairie, in order to cross near the heads of some deep ravines—we discovered a blue stripe, close in with the horizon to the west—which was by some pronounced to be no more than a cloud—by others, to be the Rocky Mountains. The hazey atmosphere soon rendered it obscure—and we were all expectation and doubt until in the afternoon, when the atmosphere cleared, and we had a distinct view of the sumit of a range of mountains—which to our great satisfaction and heart felt joy, was declared by the commanding officer to be the range of the Rocky Mountains—a high Peake was plainly to be distinguished towering above all the others as far as the sight extended—which Major Long . . . supposed distant about 60 miles. The whole range had a beautiful and sublime appearance to us, after having been so long confined to the dull uninteresting monotony of prairie country. . . .

The "high Peake" was assumed to be the "Highest Peake" of Zebulon Pike. In fact, it was then nameless, though doubtless the Indians had a name for it, and French trappers had dubbed it "Les deux Oreilles" (two ears) because of its double crest. Later travelers came to call it Long's Peak, a name that was given formal recognition in 1823 in Henry S. Tanner's *New American Atlas*. James recorded in his diary that the mountains "appear to rise abruptly from the plain and to shoot up to an astonishing altitude." The expedition's first view of the mountains was from a point a few miles east of the site of the modern city of Brush, Colorado.

On the following day the men moved another twenty-seven miles, Bell complaining of the "innumerable swarms of sand-fly or knat—the one inflames our eyes and the bite of the other causes our faces and hands to swell attended with a constant irritation and itching." Presumably these were eye gnats (*Hippelates*) and either black flies or deer flies. On this day the expedition crossed a small tributary of the South Platte that arose in higher plains to the south.

Long plotted the creek on the map he was preparing, calling it "Bijeaus Cr." after his guide, Joseph Bijeau, who had been through this country before. Neither James nor Bell mentions the creek in his report. It appears as Bijou Creek on modern maps.

James (doubtless after Say) commented on the many ant hills, astutely noting their uniform distribution—about twenty feet apart—and their east-facing entrances: "It seems highly probable, that the active little architects thus place the entrance of their edifice on the eastward side, in order to escape the direct influence of the cold mountain winds." These were colonies of the prairie mound-building ant (*Pogonomyrmex occidentalis*), a species that was not actually named and described until 1865. The even distribution of the nests is now known to be the result of the workers of each colony aggressively maintaining a feeding territory around their nest.

July 2 was a Sunday, and the expedition remained in camp, though there was little grass for the horses. Despite the explorers' rapid pace—fifty-three miles in two days—the mountains seemed as far away as ever. James made note of prairie coneflowers (*Rudbeckia columnaris*) coming into bloom. Common purslane (*Portulaca oleracea*) was abundant, "particularly in places much frequented as licks by the bisons and other animals." This is not usually regarded as a native plant, and its occurrence in 1820 so far from settlements is puzzling; Goodman and Lawson believe that James may have collected a different species.

It is interesting to reflect on some of the plants that James did *not* see as the party approached the Rockies—plants that have overrun the countryside since their unmindful introduction by humans—for example, Russian thistle, Canada and musk thistles, leafy spurge, knapweed, and cheatgrass. Of course, there were no starlings, house sparrows, Norway rats, house mice, or feral house cats. It was a pristine landscape, devoid of the innumerable irrigation ditches that now criss-cross the fields and of any roads aside from the vague trails made by bison, elk, deer, and Indians.

It was probably on this day that Seymour painted his watercolor, *Distant View of the Rocky Mountains*. As a hand-colored aquatint, it appeared as the frontispiece to the *Account*. Snowy Long's Peak is at the center, with lower ranges on each side. Bison graze on the plain before the mountains, and a few Indians stand in the left foreground (though the expedition had not encountered Indians in some days). William H. Goetzmann and William N. Goetzmann speak of it as "an epic picture," capturing the sense of "awesome space" despite its somewhat conventional style. From the profile of the mountains in Seymour's painting, it appears that it was made somewhere near the site of the present-day town of Orchard, Colorado.

On July 3, the men mounted their horses at five in the morning, hoping to reach the mountains by Independence Day. Had they headed straight toward the mountains, they might have succeeded, but the Platte turned southward, and they were committed to following its course. That day they passed the mouths of three large creeks, now called the Cache la Poudre, Big Thompson, and St. Vrain. The creeks were duly mapped by Long, but they remained unnamed except for the third, which was called Potero's (or Potera's) Creek, after a Frenchman who "is said to have been bewildered upon it." The party camped that night near the mouth of the third creek. They were near the place where Ceran St. Vrain would build a trading post in the 1830s, and not far from the site of the present town of Platteville. They were approximately due east of Long's Peak, but there was no talk of climbing that precipitous landmark. It was not until 1868 that John Wesley Powell, William N. Byers, and five others reached the top. Its altitude is now known to be 14,255 feet above sea level, roughly 9,000 feet above the campsite of Long's party.

JULY 4. EJ: We had hoped to celebrate our great national festival on the Rocky Mountains; but the day had arrived, and they were still at a distance. Being extremely impatient of any unnecessary delay, which prevented us from entering upon the

Etching in the *Account*, after Samuel Seymour, (Reproduced by permission of the Huntington Library, San Marino, California) *Distant View of the Rocky Mountains.*

examination of the mountains, we did not devote the day to rest, as had been our intention. We did not, however, forget to celebrate the anniversary of our national independence, according to the circumstances. An extra pint of maize was issued to each mess, and a small portion of whiskey distributed.

They did stop early and spent the rest of the day, according to Bell, "feasting on boiled corn soup, roasted venison and buffalo . . . Not having drank spirits for some time, the whisky tasted disagreeable."

It was probably at this site that Seymour sketched his *View of the Rocky Mountains on the Platte 50 Miles from their Base*, which appeared as an engraving in the *Account*. Several bison graze in the foreground, and in the background is the full range of the snow-capped mountains now called the Indian Peaks, with Long's Peak at the far right. Many years later, in the 1870s, Titian Peale used Sey-

Titian Peale, *Western Landscape*, 1870s. Peale based this oil painting on a sketch by Samuel Seymour, but added pronghorn antelope from one of his own sketches. (Joslyn Art Museum, Omaha, Nebraska; gift of M. Knoedler & Co., Inc., New York, N.Y.)

mour's sketch as the basis for an oil painting, adding to the foreground two pronghorn antelope from his own sketches. Peale's painting, now in the Joslyn Art Museum, Omaha, has long been attributed to Seymour.

The naturalists used the afternoon profitably. Nests and young of mockingbirds were seen in bushes by the river. Say watched prairie dogs and wondered why they selected such barren places for their villages; perhaps it was to have "an unobstructed view of the surrounding country," where predators might lurk. He collected "rattlesnakes of a particular species" that inhabited the villages. He prepared a description of them, calling the species *Crotalus tergeminus* (now *Sistrurus catenatus tergeminus*, western massasauga). This is a relatively small, mild-tempered rattlesnake that feeds

Titian Peale, sketch of scarlet gilia, probably at the campsite near present-day Brighton, Colorado, July 4, 1820. James discussed this species in his diary entry of July 3. (From the sketchbooks of Titian Ramsay Peale, Yale University Art Gallery, gift of Ramsay MacMullen, M.A.H. 1967).

mainly on small rodents. At the present time, the western massasauga is restricted to southeastern Colorado and points farther south. One wonders (here as elsewhere) if Say lost his notes concerning collection sites when his possessions were stolen by deserters. Possibly he collected the snake in the Arkansas River Valley and later reconstructed the site from memory.

James found that some of the cottonwoods in the area were "the long-leafed cotton-wood" of Lewis and Clark. The species had never been formally described, so James did so, naming it *Populus angustifolia*, narrowleaf cottonwood. He also collected "a large suffruticose species of lupine," probably *Lupinus argenteus*. He observed "the splendid and interesting Bartonia, and B. nuda in full flower." This genus had been named for Benjamin Smith Barton, who had tutored Meriwether Lewis in botany before his trip to the Pacific. Unfortunately the genus had an earlier name, *Mentzelia*, after an early German botanist. Thus these plants are now called *Mentzelia nuda*, or small white evening star. As James noted, the flowers open only in the evening or on cloudy days. These and other members of the genus are among the most striking flowers of the High Plains and foothills. Other plants noted there included a buttercup (*Ranunculus*), a beardtongue (*Penstemon*), and a figwort (*Scrophularia*).

On July 5, the explorers traveled only about ten miles, having on their right, in Bell's words, "the range of snow cap'd mountains, on the left an extensive barren prairie, almost as sterile as the deserts of Arabia." (Present-day residents of Adams County, Colorado, may well wonder at this last comment!) They camped near what French trappers called Cannon-ball Creek, "from the size and form of the stones in its bed" (modern Clear Creek). James and Peale, with two others, set out to follow the creek, tramping through fields of troublesome porcupine grass, or needle grass (*Stipa*), its "quills" penetrating "into every part of the dress with which they come in contact." They saw robins for the first time since leaving the Missouri. They forgot to take a lunch, but "Mr. Peale was fortunate enough to kill a couple of curlews, which were roasted and eaten without loss of time." They gave up without reaching the mountains, and returned at sunset with a freshly killed antelope.

On the excursion, James noted currant bushes (*Ribes*), virgin's bower (*Clematis*), blazing star (*Liatris*), water parsnip (*Sium*), wild flax (*Linum lewisii* [named by Frederick Pursh for Meriwether Lewis]), and wild buckwheat, which James called *Eriogonum sericeum* (it was later described and named *E. jamesii* by English botanist George Bentham on the basis of specimens James collected).

JULY 6. JB: At ½ past 4 o'clock in the morning, our party was on the march, delighted at the flattering prospect of arriving at the base of the mountains, the extent of our tour in this direction. . . . [A]s we advance the variety [of] objects along the mountain, change [of] scenery and of views, interest our feelings so much that we forget our fatigues. . . . [C]rossed a stream called Vermillion creek [probably present Cherry Creek, which flows through Denver]. . . . [A]bout 11 oclock we arrived within ¼ of mile of the gap between the mountains, where issued the South west branch of the river Platte, beyond which there was no possibility of advancing with horses. Our party encamped on a small plain of the river bottom which afforded good feed for our horses—a number of large cotton wood trees to shade us from the rays of the sun—& good cool water from the river— the gravell hills of the prairie, in the rear of our camp excluded the breezes from us & made the site extremely warm, set the flag on the hill & a sentinel to look out.

They had at last reached the Rocky Mountains. Bell calculated that they had traveled 568 miles since leaving Engineer Cantonment: "The Commanding officer does not intend that the party shall proceed to the source of the Platte." Presumably Long judged, from the size of the river as it left the mountains—twenty-five yards wide and averaging three feet deep—that it would have taken many days of difficult travel to reach the source, as, of course, it would have.

Secretary of War Calhoun had specifically included in his orders "an excursion . . . to the source of the river Platte." Long was therefore taking matters into his own hands in declining to pursue the South Platte through the mountains to its sources in South Park, a vast, grassy basin surrounded by towering peaks. South Park was well known to the Indians and to Canadian trappers, and James Purcell had visited the area in 1805, as mentioned in Chapter 1. Pike had entered South Park from the south, but no one had described the river between its sources and its entry onto the plains. Long might have made a major contribution by following the South Platte through its canyons, but his expedition had now been in the field for a full month, and there were many miles to go if they wished to return before autumn.

Eight
EXPLORING THE FRONT RANGE

THE EXPEDITION REMAINED CAMPED AT THE mouth of the canyon
of the South Platte for two days. Seymour took occasion to make a
watercolor painting, *View of the Chasm Through which the Platte Issues
from the Rocky Mountains*. It was reproduced, in black and white, in
the final report of the expedition. The mouth of the canyon is today
included in Waterton Canyon Recreation Area, a mecca for pic-
nickers, hikers, and fishermen. Although the area is much transformed, it is possible to look up the canyon and enjoy a view very
much like that shown in Seymour's painting. The cottonwoods
noted by Bell still provide abundant shade, and some of the trees
and bushes noted by James are still much in evidence: box elders,
oaks, junipers, and poison ivy.

At the camp, James assumed his geologist's cap and described
the formations at the rim of the mountains with considerable accuracy.

JULY 6. EJ: The woodless plain is terminated by a range
of naked and almost perpendicular rocks, visible at a distance
of several miles, and resembling a vast wall, parallel to the base
of the mountain. These rocks are sandstone. . . . They emerge
at a great angle of inclination from beneath the alluvial of the

Engraving in the *Account*, after Samuel Seymour, *View of the Chasm Through which the Platte Issues from the Rocky Mountains.*

plain, and rise abruptly to an elevation of one hundred and fifty, or two hundred feet. Passing within the first range, we found a narrow valley separating it from a second ridge of sandstone, of nearly equal elevation and apparently resting against the base of a high primitive hill beyond.

James believed that the sandstones of what we would now call the hogbacks were "originally of uniform elevation and uninterrupted continuity," but had been "broken off and thrown into an inclined or vertical position" by the uplifting of more ancient, granitic rocks: "It is difficult . . . to prevent the imagination from wandering back to that remote unascertained period, when the billows of the primeval ocean lashed the base of the Rocky Mountains, and deposited, during a succession of ages, that vast accumulation [of sedimentary rock]." For its time, this was a perceptive description of the formation of the Rockies and of the hogbacks.

Nowadays we can provide a rough time scale for these events and can supply names for the formations, but most of James's com-

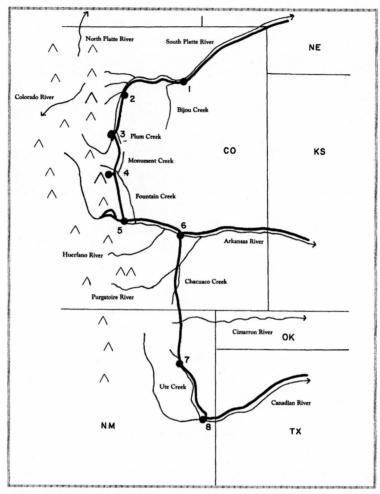

Route of the Long Expedition through Colorado and New Mexico, with major landmarks indicated: (1) first view of Long's Peak, near present Brush, Colorado, June 30; (2) camp opposite Potera's (St. Vrain) Creek, near present Platteville, Colorado, with Long's Peak to the west, July 3; (3) camp on the South Platte where it leaves the mountains, present Waterton Canyon, just southwest of Denver, July 6; (4) Edwin James and two others climb Pike's Peak, July 13–15; (5) camp on the Arkansas near the mountains, July 16–19; (6) expedition splits, near present Rocky Ford, Colorado, July 22–24; (7) arrival at Ute Creek, near Gladstone, New Mexico, July 30; (8) arrival at the Canadian River, near Logan, New Mexico, August 5.

ments are still applicable. Modern geologists agree that the sedimentary rocks were laid down horizontally, beneath shallow seas, and
were tilted upward by the slow elevation of more ancient granitic
and metamorphic rocks; most of the sedimentary rocks that overlaid
the much older rocks have been eroded away, leaving parallel ridges
of sandstone and limestone. The first volume of Charles Lyell's rev-
olutionary *Principles of Geology* was not published until 1830, and
prior to that time there was little appreciation of the massive forces
and the vast amount of time required to form the landscapes of today.
James was clearly a forward-looking geologist.

The men of the expedition were camped between James's "first
range" (Dakota Sandstone, in modern terminology) and his "second
ridge" (Lyons and Fountain Formations), the latter resting against
the "high primitive hill beyond." The ridges, he noted, had been
"cut through by the bed of the Platte, and all the larger streams in
their descent to the plains." "From our camp, we had expected to
be able to ascend the most distant summits then in sight, and return
the same evening." As is always the case, the mountains were found
to have greater depth than they appeared from their base.

On the hogbacks, James collected a wild geranium and "the
beautiful calochortus." This was a sego lily, or mariposa lily (*Calochortus gunnisoni*), a plant still abundant in the area. James misidentified the plant as *C. elegans*, and it was not until 1871 that it was
recognized as a different species and named for Captain John W.
Gunnison, a notable western explorer who fell to the Paiutes in
1853. The bulbs of sego lilies were often harvested as food by Native
Americans, and in 1848 the Mormon settlers of Utah relied on them
after their crops had been devastated by crickets. *C. nuttallii* is now
the state flower of Utah.

James noted "an uncommonly large and beautiful buck deer
[that sprang] out from the bushes, and stood gazing at us, until he
received in his side the ball, which brought him instantly to the
earth." So much for beauty!

Say discovered and described two specimens belonging to a

group of rather grotesque arachnids usually called sun spiders, wind scorpions, or solpugids. One he named *Galeodes pallipes*; the other, *G. subulata*. It was later discovered that one was a female and the other a male of the same species, now called *Eremobates pallipes*. This was the first record of these unusual arthropods from North America. We now know that there are numerous species in the western states. Sun spiders are agile, pale-colored predators, an inch or two in length, that seize their prey in pincers that seem too large for their slender bodies. They are most active at night, and despite their fierce appearance they are not poisonous.

Say also described a previously unknown wren that was "often seen hopping about on the branches . . . and trunks [of juniper]" at the campsite. He named it *Troglodytes obsoleta* (now *Salpinctes obsoletus*, rock wren). Say said nothing about the song of these birds, which is among the most exuberant and varied of any Rocky Mountain bird. The species deserves a better name than *obsoleta*, which means "worn out" in Latin. Say described the plumage as "of a much more sombre hue" than that of the Carolina wren, which may explain why he chose that name.

On the morning of July 7, two groups set off into the nearby mountains on foot. James and Peale, with two riflemen, explored the hills on the north side of the river. Since they were camped on the south side, this required that they cross the river, some twenty-five yards wide and with a current so strong "as to render it impossible for a man to keep on his feet." So one of the riflemen swam the stream with the end of a rope in his teeth, and then made the rope fast on the other side so that others could wade the stream while holding on to the rope. They found the south-facing slopes covered with prickly pear cacti and yuccas, "with here and there a stinted oak or juniper, and so steep that great exertion as well as the utmost caution, are necessary in ascending." A few "interesting insects and plants" were found, but the slopes were otherwise "the abode of few living things, either animal or vegetable." After an arduous climb, they were able to look down on the river and see the forks of the

South Platte—"Two Forks," where the Denver Water Board has in recent years yearned to build a dam. The descent in the afternoon was as laborious as the ascent, but the men did find ripe currants and "a few large and delicious raspberries." These were Boulder raspberries (*Rubus deliciosus*), described and named by John Torrey on the basis of specimens collected by James.

When they were several miles into the mountains, one of the riflemen became too ill to proceed, and Peale set out for camp to find help. Soon after he left, a large bear approached the others in the group, but it was frightened off with some shots. The sick man soon recovered, and the party arrived back at camp in the evening only to find that others had become ill.

Bell and Say, who had ascended the hills on the south side of the river, also experienced many difficulties. They were rewarded by a close view of elk, a feast of currants, and a broad view of the mountains ahead as well as the prairie they had left behind. They attempted to cross the river to find an easier way back; but the current was too strong, the water cold, and the bottom covered with sharp rocks, so they were forced to scramble through rocks and bushes along the mountainside. They were also ill upon return.

> EJ: The sickness experienced by almost all the party was probably occasioned by eating of currants, which were abundant about the camp. It is not to be supposed this illness was caused by any very active deleterious quality of the fruit, but that the stomach, by long disuse, had in great measure lost the power of digesting vegetable matter. Several continued unwell during the night.

Bell was more graphic, describing the "violent pains in the head, breast & limbs, in some cases attended with vomiting & surging." But, he added, "a dose of calomel & jallup administered, soon gave relief—we eat no more currants." Perhaps their illness was caused, or aggravated, by the rigors of their first climb into the mountains: they may have been experiencing a touch of altitude sickness.

Despite his description of the bleakness and hazards of his trip into the mountains, James made note of several plants he had found there. They included ninebark (*Spiraea opulifolia*, now *Physocarpus monogynus*); common hops (*Humulus lupulus*); bearberry, or kinnikinnik (*Arctostaphylos uva-ursi*); shooting star (*Dodecatheon pulchellum*); and common sarsaparilla (*Aralia nudicaulis*). An "undescribed acer" was Rocky Mountain maple (*Acer glabrum*), described several years later by Torrey on the basis of specimens collected by James.

The next day the expedition proceeded south. They were late in starting, as they waited for James to retrieve a collection of plants he had lost during his climb. The men followed a small stream through a valley to the east of the hogbacks. Occasionally, wrote Bell, they "passed an opening made by the outlet of a rivulet from the mountains, where we had new and very interesting views of insolated masses of rock . . . of singular colour and formation, the whole scenery truly picturesque & romantic." Many of the same formations that excited the explorers may be seen today in Roxborough State Park, just a few miles southwest of Denver. Visitors to the park pass through a break in a hogback of Dakota Sandstone (doubtless one of the openings noted by Bell) and enter a narrow north–south valley filled with hillocks of whitish Lyons Sandstone and massive tilted slabs of the Fountain Formation, colored deep red by iron compounds. To James, some of these rock masses bore "a striking resemblance to colossal ruins."

Bearing southeast, the men crossed "an inconsiderable ridge" and entered the valley of a north-flowing tributary of the South Platte, which they called Defile Creek (now Plum Creek). The mountains were to their right, rolling hills to their left. They covered fourteen and a half miles that day, camping about ten miles south of the site of the present-day town of Sedalia on the west fork of Plum Creek. They remained there for two nights, July 9 being a

Titian Peale, sketch of tilted rocks paralleling the foot of the Rockies, near the campsite of July 6–7, 1820. These deep reddish-brown rocks date from the Paleozoic. (From the sketchbooks of Titian Peale, Yale University Art Gallery, gift of Ramsay MacMullen, M.A.H. 1967)

Sunday. Here James assumed his role of physician and bled many of the explorers.

> JULY 9. JB: [C]an it be, our great altitude above tide water, that blood letting has of late become so necessary? Until that operation is performed on several of the men the blood in their faces seems almost ready to break thro' the skin, their eyes inflamed & violent pains in the head.

In the early nineteenth century, blood-letting was considered desirable to counteract the effects of high altitude, though it would have weakened men who were exhausted from travel and only just recovering from an illness they attributed to eating currants. Bell commented on their diet (as he often did); it consisted of "wild meats & about 3 oz of hard bread pr. day."

EJ: In that part of Defile creek, near which we encamped, are numerous dams, thrown across by the beaver, causing it to appear rather like a succession of ponds than a continued stream. As we ascended farther towards the mountains, we found the works of these animals still more frequent. The small willows and cotton-wood trees, which are here in considerable numbers, afford them their most favorite food.

Although Sunday was supposedly a day of rest, the men were particularly busy. In the morning, Long and one of his men climbed a hill (Dawson Butte) to the east of the valley.

EJ: The ascent of the hill is steep and rugged; horizontal strata of sandstone and coarse conglomerate are exposed on its sides, and the summit is capped by a thin stratum of compact sandstone surmounted by a bed of greenstone [volcanic rock]. The loose and splintery fragments of this rock sometimes cover the surface and make a clinking noise under the feet like fragments of pottery.

The presence of volcanic deposits is unexpected; evidently, they represent ash from an area of past vulcanism to the southwest, near present Canon City.

From the top of the butte, Long got his first glimpse of the peak that Pike had found, and determinations of latitude made it certain that the peak the explorers had passed a few days earlier (now Long's Peak) was in fact previously unknown (or at least unreported). Surely they had already suspected this, since Pike had reported his peak as being in the drainage of the Arkansas rather than that of the Platte.

On the same day, James, Peale, and Seymour followed a stream for a distance toward the mountains, and James recorded the rock formations "beautifully exposed" by the stream cut. Starting in an area of horizontal sandstone, they traversed a hogback of gray rocks (Dakota Sandstone), tilted toward the mountains. This was followed by "lofty and detached columns of sandstone of a reddish or deep brown colour." Some of them were too tall and steep to climb, but they did climb one of them. Finally, they entered a region of "coarse white pudding-stone, or conglomerate and sandstone of a deep red colour, alternating with each other." They were now in rocks of the Lyons and Fountain Formations, which provide the most colorful and spectacular formations along the Front Range. Having postulated the origin of these bands of rocks upon his first arrival at the mountains, James was surely aware that he was crossing a transect involving vast stretches of time. Indeed he was: we now know that Dakota Sandstone dates from the late Mesozoic, perhaps 100 million years ago, and that rocks of the Fountain Formation were laid down in the late Paleozoic, roughly 300 million years ago. Beyond the last of the sedimentary rocks rose granite "in immense mountain masses" undoubtedly extending "far to the west." The creek the men were following, continued James, poured "down the side of this granitic mountain through a deep, inaccessible chasm, forming a continued cascade of several hundred feet."

The men were at a place now called Perry Park, where visitors may vicariously experience some of the excitement that James and his companions must have felt as they examined these colorful formations so many years ago. Here tall and often vertical and jagged rock formations seem thrown about as if by a mad creator. Today Perry Park is filled with houses, a golf course, and a small lake formed by damming the creek.

At the base of the mountains, the men discovered a brine spring, about which were skulls of bighorn sheep and bones of elk, bison, and other animals. "A beautiful species of pigeon" was shot near the mountains; it was purple and green, with a pale band along the tail. Say provided a description of this elegant bird, the band-tailed pigeon. He named it *Columba fasciata*, a name that still stands. These are birds of coniferous woodlands and oak scrublands, where they gorge on acorns, juniper berries, and pine seeds. The birds can still be seen frequently in the area where they were first discovered.

We now know that they range throughout the southern Rockies and along the Pacific coast.

Say also described another large bird from this site, calling it the dusky grouse and naming it *Tetrao obscurus*. This is the blue grouse, now called *Dendragapus obscurus*. "When this bird flew," he wrote, "it uttered a cackling noise a little like that of the domestic fowl." In fact, the blue grouse was the first bird ever reported from what is now Colorado, having been discovered in 1776 by Fathers Francisco Domingues and Silvestre Vélez de Escalante as they traversed the western part of the state on the way (they hoped) to California. They spoke of it as "a kind of chicken" and "exceedingly palatable." But it remained for Say to provide a formal name for the grouse, now a favorite game bird in the West. In the mating season, the male establishes a territorial perch on a horizontal branch of a pine, raises his tail, and inflates air sacs on the sides of his neck, producing a series of owl-like hoots.

The morning of July 10 was so cool that the explorers stood around the campfire warming themselves before mounting their horses a few minutes before five. The hunters were sent out and returned in a few hours with a bison and an antelope. They also saw a "white bear," as the grizzly was then often called. The expedition had now ascended the valley of West Plum Creek to an elevation of about 7,000 feet. Much of the country was wooded with pines and oaks. The oaks, James noted, resembled *Quercus banisteri* of André Michaux; they were in reality *Q. gambelii*, a species not described until 1848 by Thomas Nuttall. The pines were identified as red pines (*Pinus resinosa*); they were in fact ponderosas (*P. ponderosa*), a species yet without a name. It was David Douglas who, in 1836, sent specimens to England and suggested the name *ponderosa*. Today, the landscape through which the expedition was now passing remains thinly settled, still dominated by Gambel oaks and ponderosa pines.

By eight o'clock, the expedition passed over a low divide and "bid adieu to the waters of the Platte." Before them lay a pond (Palmer Lake) beyond which a stream flowed southward. Just past

Engraving in the *Account*, after Samuel Seymour, *View of the Insulated Table Lands at the Foot of the Rocky Mountains*. In the foreground, the expedition passes in single file.

the pond was a spectacular rock formation, with a rooflike caprock, columns, porticos, and even an arch suggesting a spacious entryway, the whole resembling "an extensive edifice in ruins." While the men made a lunch stop, Seymour prepared a sketch. Today this formation can easily be seen from the road between the villages of Palmer Lake and Monument. Long called it Castle Rock, and the south-flowing stream they were about to follow he named Castle Rock Creek. It is now called Elephant Rock (the name Castle Rock is now applied to a butte some miles to the north, and to the city at its base). Long's Castle Rock Creek is now called Monument Creek. As Richard Beidleman has pointed out, the countryside immediately around Elephant Rock has changed little since 1820, aside from encroachments of suburbia and a denser growth of trees. Some of the Douglas-firs that Seymour sketched adjacent to the rock are still thriving. Some of them are known to be as much as 350 years old.

During the pause near Elephant Rock, James made a short hike into the mountains and collected "a large species of columbine . . . heretofore unknown to the flora of the United States, to which it forms a splendid acquisition. . . . [I]t may receive the name of aqui-legia caerulea." This is blue columbine, one of the most cherished of Rocky Mountain flowers. James provided a formal description, remarking that "it inhabits shady woods of pine and spruce within the mountains, rising sometimes to the height of three feet." In 1891, Colorado schoolchildren voted blue columbine the state flower of Colorado, and a few years later a bill to that effect was passed by the legislature and signed into law by the governor.

Other plants seen or collected that day included quaking aspen (*Populus tremuloides*), northern green orchid (*Orchis dilatata*, probably *Habenaria hyperborea*), and one-sided wintergreen (*Pyrola secunda*). James reported both "black and hemlock-spruce, (Abies *nigra* and A. *canadensis*)." These are trees of more eastern and northern parts of the continent, and do not occur in Colorado. Doubtless he found not those species but Douglas-firs, first recognized as a distinct species by David Douglas on the Pacific coast in 1825. Or he may have seen white firs or Colorado blue spruce, neither of which was recognized as distinct species until the 1860s.

As the men lunched near Elephant Rock, James noted among the grasses "a small campanula" that he recognized as similar to the common bluebell of Europe. This was a harebell (*Campanula*); ac-cording to George Goodman and Cheryl Lawson probably C. *parryi*, described by Asa Gray many years later. No bison had been seen for several days, but several were spotted in the distance, and hunters were sent out to take them. The naturalists were treated to "the animated spectacle of a bison hunt," and the hunters returned with "their horses loaded with the meat."

JULY 10. JB: We are delighted with our first entrance of what we may now call Arkansas country—cool water from the mountains, numberless beaver dams & lodges. Naturalists find new inhabitants, the botanist is at [a] loss which new plant he will first take in hand—the geologist grand subjects for specu-lation—the geographer & topographer all have subjects for ob-servation. . . . [I]n the ravines is to be seen fine growth of pine timber, a great relief to the eye, after having been so long ac-customed to see none but cotton-wood—about 6 p.m. halted and encamped for the night, at a place where we was almost surrounded by groves of pine trees, soil sandy—the hunters came up at 12 oclock having killed a buffalo & seen a white bear—they [went] out again after we encamped, killed an an-telope.

JULY 11. JB: Marched before 5 oclock a.m. continuing our course down the margin of the creek—the number of rivulets from the mountains & one from the prairie has increased the size of the creek 8 to 10 yards in width, as many inches deep, and bed gravel—halted to refresh at 10 oclock—a high snow caped mountain in sight [Pike's Peak] which seems to be within the range, along the base of which we are travelling—the weather cool & pleasant, because of the sun being obscured by the clouds resting as it were, upon the mountains. We pro-ceeded at 1 p.m. leaving the creek to our left, taking a course more along the base of the mountain, which appeared to turn off considerably to the right. The travelling was very fatiguing for the horses, being obliged to cross a number of deep ravines, where we were obliged to dismount in descending and ascend-ing their banks or sides which are very abrupt—passed a valley which appeared to extend a great distance between the ranges of the mountains, into which led a number of well beatten Indian traces. . . .

The expedition had passed through the site of the campus of the United States Air Force Academy and through the western part of modern metropolitan Colorado Springs. The valley they had seen to the west was that of Fountain Creek, its Indian trails now replaced by U.S. Highway 24. Robins were common, and the men saw a "jerboa" (doubtless a kangaroo rat). Once again, "many fine plants were collected." One of these was prince's plume, a showy member of the mustard family, which James described in a footnote, calling it *Stanleya integrifolia*. Thomas Nuttall had described the genus, naming it for Lord Edward Stanley, president of the Linnean Society of London. James's name is now applied to a subspecies of *S. pinnata*, which Frederick Pursh had described in 1814.

In the evening the explorers realized that they had passed Pike's Peak, which had been shielded from their view by Cheyenne Mountain (the bowels of which today contain NORAD [North American Air Defense Command]). So they stopped and camped. Near the camp, James noted the presence of "the great shrubby cactus," with large purple blossoms and a "terrific armature of thorns." This was cholla, or candelabra cactus (*Opuntia imbricata*), which they would see more of later when they proceeded south. Another interesting plant was buffalo gourd, which James called *Cucumeris perennis* (now *Cucurbita foetidissima*). Peale sketched a gourd plant, which provided "by its deep green, a striking contrast to the general aspect of the regions it inhabits, which are exceedingly naked and barren." Long collected the gourd's seeds, which were later planted in the garden of the University of Pennsylvania, where they grew successfully. Other plants collected were death camas (*Zigadenus elegans*) and butterfly weed, an orange-blossomed milkweed (*Asclepias tuberosa*). The latter still enlivens the edges of the greens at the golf course of the Broadmoor Hotel.

Since Long and his men had hoped to explore the "High Peake" in more detail, on the following day they retraced their steps northward and found a place to camp on Fountain Creek, just south of present Colorado Springs. They had a fine view of the mountain; "all but the upper part was visible, with patches of snow extending down to the commencement of the woody region." Hunters were

Titian Peale, sketch of buffalo gourd, just south of present-day Colorado Springs, July 12, 1820. (From the sketchbooks of Titian Ramsay Peale, Yale University Art Gallery, gift of Ramsay MacMullen, M.A.H. 1967)

dispatched and soon brought in plenty of game. Bell once again became the expedition's gastronomic observer.

> JULY 12. JB: We [are] again feasting on abundance of choice meat, around us is exhibited a curious and interesting sight—on one side is arranged a sparerib of a buffalo, the head of a deer, 8 or 10 pieces of buffalo meat of different sizes, all supported to the heat of the fire on small sticks—the opposite side of the fire is decorated in the same manner, except in place of the rib and head, is two saddles of venison, suspended from the ends of short stakes, drove oblikely into the ground—what a sight! what a feast for an epicure!

But Bell was upset later when a storm in the mountains caused the stream to rise and become filled with bison dung, creating "a most intolerable stench." The men drew water in kettles, let it settle, and then skimmed off the dung. "This clarified, we used it to drink and boil our meat in, but the disagreeable smell remained in our soup." James was equally distressed: "[W]hen the soup was brought to our tent, the flavour of the cow-yard was found so prevalent, and the meat so filled with sand, that very little could be eaten."

It was at or near this campsite that Seymour painted one of his most effective watercolors: *View of James* [Pike's] *Peak in the Rain.* This painting was not reproduced in the *Account,* and for many years was incorrectly attributed to William McMurtrie, an artist who visited Colorado in 1854. Recent research has established that it is Seymour's work, the first rendition of what became a major landmark for overlanders. The painting is now in Boston's Museum of Fine Arts and is reproduced in color in Patricia Trenton and P. H. Hassrick's book *The Rocky Mountains: A Vision for Artists in the Nineteenth Century.*

At the campground, Say meanwhile noted "a very pretty little bird . . . hopping about in the low trees or bushes, singing sweetly, somewhat in the manner of the American goldfinch." He recognized it as previously unknown and supplied a description and a name, *Fringilla psaltria.* The species is now known as the lesser goldfinch (*Carduelis psaltria*). Another attractive member of the finch family he called the crimson-necked finch, providing the scientific name *F. frontalis.* This species is now called the house finch; its current name is *Carpodacus mexicanus frontalis.* Today this is one of the common birds of settled areas throughout the West. It was introduced into the eastern states in the 1940s and is well established in many places there.

Say was also busy collecting insects, and since no one had ever collected them in this part of the country, the majority of his specimens represented previously unknown species. As was his custom, he did not include descriptions in the report of the expedition, but did so in several later publications. A perusal of these reveals some thirty species collected "near the Rocky Mountains," without further

Samuel Seymour, *View of James Peak in the Rain*, 1825. Done in ink and water-color, this is believed to be the first illustration of Pike's Peak. (M. & M. Karolik Collection, courtesy of Museum of Fine Arts, Boston)

information. It would not do to list all of them here, but several are worthy of note. It may have been at this camp that Say collected *Mutilla* (now *Dasymutilla*) *quadriguttata*, an elegant wingless parasitic wasp, or "velvet ant," of mostly reddish coloration and with four pale spots on its abdomen. And perhaps it was here that he collected an ornate parasitic bee (*Epeolus* [now *Triepeolus*] *quadrifasciatus*), and a small black and red digger wasp (*Astata bicolor*).

Beetles were favorites of Say and were doubtless easier to catch and preserve than were butterflies and some other insects. Some of the local bushes "were loaded" with blister beetles, which he named for his friend Thomas Nuttall, calling the species *Lytta nuttalli*. This is a beautiful green and purple species, illustrated in color in the first volume of *American Entomology*. Also illustrated was a brilliantly colored, soft-bodied beetle he called *Lycus* (now *Lysostomus*) *san-guinipennis* (the species name is Latin for "blood-colored wings").

Beetles of this group are known to be extremely distasteful to birds, a fact they advertise with their vivid colors. They are mimicked by quite a variety of more palatable insects, including moths, flies, and several kinds of beetles.

The expedition remained camped on Fountain Creek for three days, allowing time for some of the group to attempt a climb of the "High Peake." Before dawn on July 13, James and four others headed for the mountain. For the first day, they were accompanied by William Swift and the guide, Joseph Bijeau. It was Swift's job to determine the height of the mountain. He calculated a height of 8,500 feet above the base, which he thought to be about 3,000 feet above sea level. Thus had he known that the base was actually about 6,000 feet in elevation, he would have estimated a total elevation of 14,500 feet—not far from the presently known elevation of 14,110 feet. Pike had estimated the elevation as 18,541 feet and had considered the mountain unclimbable.

Near the base of the mountain, at about eleven in the morning, a place was found to leave the horses, and the group proceeded up the valley on foot. Around noon, they arrived at a bubbling spring, "a large and beautiful fountain of water, cool and transparent, and aerated with carbonic acid." The waters left a basin of "snowy whiteness, and large enough to contain three or four hundred gallons . . . constantly overflowing." The spring produced a rumbling noise and was found to be of "grateful taste [with] the exhilarating effect of the most highly aerated artificial mineral waters."

A second spring nearby contained a smaller bubbling pool, but discharged no flow. James found the water temperature to be sixty-three degrees Fahrenheit in the larger pool, and sixty-seven degrees in the smaller one. (These are now called Manitou Springs, the site of a town of several thousand people.) The spring contained beads, shells, and other objects thrown in by Indians as an offering. Bijeau reported that French traders sometimes took objects from the spring and traded them back to the Indians for furs.

At the springs, Swift and Bijeau were left behind, and James

and two others, Zachariah Wilson and Joseph Verplank, headed up
the mountain. Each had "a small blanket, ten or twelve pounds of
bison meat, three gills of parched corn meal, and small kettle." Some
food was left in a tree near the springs, for use upon returning.

Since the first ascent of Pike's Peak is one of the classics of
western history, I shall, for the most part, let James describe it in his
own words. He and his two companions were not only the first persons to climb Pike's Peak, but the first to climb any of the fifty or so
14,000-foot peaks in Colorado. That a young and energetic botanist
made the first climb is especially fitting, for here was a new world—
an alpine flora almost wholly unknown to science. It was a climb
not without adventure.

EJ: The ascending party found the surface in many places
covered with such quantities of this loose and crumbled granite,
rolling from under their feet, as rendered the ascent extremely
difficult. We now began to credit the assertions of the guide,
who had conducted us to the foot of the peak, and there left us,
with the assurance that the whole of the mountain to its summit was covered with loose sand and gravel; so that, though
many attempts had been made by the Indians and by hunters to
ascend it, none had ever proved successful. We passed several of
these tracts, not without some apprehension for our lives, as
there was danger, when the foothold was once lost, of sliding
down, and being thrown over precipices. After labouring with
extreme fatigue over about two miles, in which several of these
dangerous places occurred, we halted at sunset in a small cluster
of fir trees. We could not, however, find a place of even ground
large enough to lie down upon, and were under the necessity of
securing ourselves from rolling into the brook near which we
encamped by means of a pole placed against two trees. In this
situation, we passed an uneasy night; and though the mercury
fell only to 54°, we felt some inconvenience from cold.

On the morning of the 14th, as soon as daylight appeared,

having suspended in a tree our blankets, all our provisions, except about three pounds of bison's flesh, and whatever articles of clothing could be dispensed with, we continued our ascent, hoping to be able to reach the summit of the peak, and return to the same camp in the evening. After passing about half a mile of rugged and difficult travelling, like that of the preceding day, we crossed a deep chasm, opening towards the bed of the small stream we had hitherto ascended; and following the summit of the ridge between these, found the way less difficult and dangerous.

Having passed a level tract of several acres covered with [trees], we arrived at a small stream running towards the south, nearly parallel to the base of the conic part of the mountain which forms the summit of the peak. From this spot we could distinctly see almost the whole of the peak; its lower half thinly clad with pines, junipers, and other evergreen trees; the upper, a naked conic pile of yellowish rocks, surmounted here and there with broad patches of snow. But the summit appeared so distant, and the ascent so steep, that we began to despair of accomplishing the ascent and returning the same day.

About the small stream before mentioned, we saw an undescribed white-flowered species of caltha, some pediculariae, the shrubby cinque-foil (potentilla fruticosa, Ph.) and many alpine plants.

The *Caltha* was undoubtedly *leptosepala*, white marsh marigold; the species had been described just two years earlier, by Swiss botanist Augustin de Candolle. This is an abundant plant in wet, subalpine meadows and was used as a potherb by the Indians. Pediculariae are louseworts, of which there are several species in the Colorado mountains. They were once supposed to cause sheep to become lousy when they fed on the plants, hence the name (*pedicularis* is the Latin word for "louse"). As they ascended, James noted many stonecrops (*Sedum lanceolatum*). The trees gradually became smaller, and at noon they arrived at timberline; still, "a greater part of the whole elevation of the mountain seemed before us."

> EJ: A little above the point where the timber disappears entirely, commences a region of astonishing beauty, and of great interest on account of its productions. The intervals of soil are sometimes extensive, and covered with a carpet of low but brilliantly-flowering alpine plants. Most of these have either matted procumbent stems, or such as, including the flower, rarely rise more than an inch in height. In many of them the flower is the most conspicuous and the largest part of the plant, and in all the colouring is astonishingly brilliant.
>
> A deep blue is the prevailing colour among these flowers; and the pentstemon erianthera [a beard-tongue], the mountain columbine (aquilegia coerulea), and other plants common to less elevated districts, were much more intensely coloured than in ordinary situations. . . .

By two o'clock, James and Verplank were exhausted and paused to rest. Wilson had been left behind and could not be seen anywhere. Shouts and a rifle shot failed to bring a response, so the two lunched and trusted that Wilson would catch up with them.

> EJ: Here, as we were sitting at our dinner, we observed several small animals, nearly the size of the common gray squirrel, but shorter, and more clumsily built. They were of a dark gray colour, inclining to brown, with a short thick head, and erect rounded ears. In habits and appearance, they resemble the prairie dog, and are believed to be a species of the same genus. The mouth of their burrow is usually placed under the projection of a rock; and near there the party afterwards saw several of the little animals watching their approach, and uttering all the time a shrill note, somewhat like that of the ground squirrel. Several attempts were made to procure a specimen of this an-

imal, but always without success, as we had no guns but such
as carried a heavy ball.

These animals were, of course, pikas. Had James and his companions
taken a specimen, they might have brought it to Say to describe and
thereby gain credit for making the species known to science. As it
was, it was not until 1828 that the pika was formally described by
John Richardson, who had found it in Alberta. James was incorrect
in suggesting a similarity to prairie dogs; pikas are much more closely
related to rabbits.

> EJ: After sitting about half an hour, we found ourselves
> somewhat refreshed, but much benumbed with cold. We now
> found it would be impossible to reach the summit of the moun-
> tain, and return to our camp of the preceding night, during that
> part of the day which remained; but as we could not persuade
> ourselves to turn back, after having so nearly accomplished the
> ascent, we resolved to take our chance of spending the night
> on whatever part of the mountain it might overtake us. Wilson
> had not yet been seen; but as no time could be lost, we resolved
> to go as soon as possible to the top of the peak, and to look for
> him on our return. We met, as we proceeded, such numbers of
> unknown and interesting plants, as to occasion much delay in
> collecting; and were under the mortifying necessity of passing
> by numbers we saw in situations difficult of access.

One can but imagine the frustration that James must have experienced, knowing he had but a few hours to sample so many novel
plants. In their book *Land Above the Trees*, Ann Zwinger and Beatrice Willard point out that the Rockies have more than 300 alpine
plant species, three times as many as the mountains of the eastern
states. James had time to collect only a few. These he succeeded in
preserving until he returned east, and most were turned over to John

Blue columbine, now Colorado's state flower. It was discovered by Edwin James near Palmer Lake, Colorado.

Torrey for description. Torrey's article was titled "Descriptions of Some New or Rare Plants from the Rocky Mountains Collected in July 1820 by Dr. Edwin James."

Among the new plants collected by James and described by Torrey were dwarf alpine clover (*Trifolium nanum*), alpine primrose (*Primula angustifolia*), rock jasmine (*Androsace carinata*), alpine blue-bells (*Mertensia alpina*), western yellow paintbrush (*Castilleja occidentalis*), alumroot (*Heuchera bracteata*), and James's saxifrage (*Telesonix jamesii*). English botanist George Bentham later added snowlover (*Chionophila jamesii*). It was by no means a bad haul for a few frantic hours on the tundra.

. At four o'clock, James and Verplank reached the summit, where Wilson managed to catch up with them. There was a panorama of snow-capped mountains from the northwest to the southwest, and directly to the west the watershed of the Arkansas River

Western yellow paintbrush, one of many plants discovered by James on his climb of Pike's Peak.

could be seen. To the north lay what may have been part of the watershed of the South Platte, and to the east a broad expanse of the plains.

> EJ: The weather was calm and clear while the detachment remained on the peak; but we were surprised to observe the air in every direction filled with such clouds of grasshoppers, as partially to obscure the day. They had been seen in vast numbers about all the higher parts of the mountain, and many had fallen upon the snow and perished. It is, perhaps, difficult to assign the cause which induces these insects to ascend to those highly elevated regions of the atmosphere. Possibly they may have undertaken migrations to some remote district; but there appears not the least uniformity in the direction of their movements. They extended upwards from the summit of the mountain to the utmost limit of vision; and as the sun shown brightly,

they could be seen by the glittering of their wings, at a very considerable distance.

Beyond doubt, these were migrating swarms of the now extinct Rocky Mountain locust (*Melanoplus mexicanus*). Remains of these grasshoppers can still be found frozen in glaciers on several peaks in the Rockies, though the last living specimen was seen in 1902. These insects once arrived in great clouds over farms in the Great Plains, feeding on almost everything in their path, even the bark of trees and clothing on the line. Having consumed everything edible, they flew off in search of greener pastures; those that flew over the mountains were presumably looking for a better food supply. Why the species abruptly became extinct remains a puzzle.

EJ: About all the woodless parts of the mountain, and particularly on the summit, numerous tracks were seen, resembling those of the common deer, but most probably have been those of the animal called the big horn. The skulls and horns of these animals we had repeatedly seen near the licks and saline springs at the foot of the mountain, but they are known to resort principally about the most elevated and inaccessible places.

The party remained on the summit only about half an hour; in this time the mercury fell to 42°, the thermometer hanging against the side of a rock, which in all the early part of the day had been exposed to the direct rays of the sun. At the encampment of the main body in the plains, a corresponding thermometer stood in the middle of the day at 96°, and did not fall below 80° until a late hour in the evening. . . .

At about five in the afternoon the party began to descend, and a little before sunset arrived at the commencement of the timber; but before we reached the small stream at the bottom of the first descent, we perceived we had missed our way. It was now become so dark as to render an attempt to proceed extremely hazardous; and as the only alternative, we kindled a fire, and laid ourselves down upon the first spot of level ground we could find. We had neither provisions nor blankets; and our clothing was by no means suitable for passing the night in so bleak and inhospitable a situation. We could not, however, proceed without imminent danger from precipices; and by the aid of a good fire, and no ordinary degree of fatigue, found ourselves able to sleep during a greater part of the night.

15th. At day break on the following morning, the thermometer stood at 38°. As we had few comforts to leave, we quitted our camp as soon as the light was sufficient to enable us to proceed. We had travelled about three hours when we discovered a dense column of smoke rising from a deep ravine on the left hand. As we concluded this could be no other than the smoke of the encampment where we had left our blankets and provisions, we descended directly toward it. The fire had spread and burnt extensively among the leaves, dry grass, and small timber, and was now raging over an extent of several acres. This created some apprehension, lest the smoke might attract the notice of any Indians who should be at that time in the neighbourhood, and who might be tempted by the weakness of the party to offer some molestation.

The men discovered that most of their provisions had been destroyed by the fire, but they managed "a beggarly breakfast" from the remains. They descended by a different route from the one they had ascended; there was less crumbly granite, but an abundance of yuccas and cacti. They arrived in the early afternoon at the springs, where they "indulged freely in the use of its highly aerated and exhilarating waters." The men who had been left at the horse camp about a mile below the springs had killed several deer, so there was plenty to eat. After feasting, the men mounted their horses and went on to the main camp, where they arrived a little after dark. The trip had been completed in the three days allotted.

James misidentified many of the coniferous trees he encountered in the mountains, listing balsam fir and several species of spruce that do not in fact occur there. It was not until some years later that the firs and spruces of the Rockies were properly classified. James did recognize as undescribed a conifer of the white pine group, with five needles in a cluster. He provided a description and named it *Pinus flexilis*; we now call it limber pine. He noted an abundance of scouring rushes, or horsetails (*Equisetum*), growing along Boiling Spring (now Fountain) Creek, "eaten avidly by horses." Along the stream, the naturalists saw many birds, including robins, towhees, Lewis's woodpeckers, red-headed woodpeckers, yellow-breasted chats, kestrels, wrens, and mockingbirds. "Orbicular lizards" (horned lizards, *Phrynosoma*) were also seen.

Besides his collections and natural history observations, James was able to provide Long with descriptions of the mountains and valleys as seen from the summit. This enabled him to rough them in on his map with respect to the ranges as well as the drainages of the upper Arkansas and South Platte Rivers. Nowadays, when both a road and a cog railway make the ascent of Pike's Peak a simple matter, it is difficult to appreciate the difficulties and the excitement experienced by James and his companions. They had made an epic climb, and Long decided to name the mountain James Peak. However, the name did not stick, and later generations have called the peak by the name of its discoverer rather than that of the first to climb it. A different peak was later named for James. It is west of Denver, not quite as high as Pike's Peak but still a substantial landmark at 13,294 feet. Not far away is Parry Peak, named for another notable botanist. (Charles Parry roamed the Rockies for plants from 1861 to about 1875, and when on Pike's Peak in 1862 collected what later would become Colorado's state tree, blue spruce.) Only about ten miles from James and Parry Peaks is a peak named for George Engelmann, and another ten miles, more or less, twin peaks named for John Torrey and Asa Gray. They are fitting monuments to five botanists who did much to make the plant life of the Rockies known to science.

Jamesia, or waxflower. It was described by John Torrey and Asa Gray on the basis of specimens collected by James and was named for him.

But perhaps the greatest tribute to James's accomplishments was the naming of *Jamesia* by Torrey and Gray in 1840. This is a unique genus of plants, confined to rocky slopes and crevices in the West. It is a member of the hydrangea family and is often called wild hydrangea, cliffbush, or waxflower. It is a handsome shrub, bearing in June clusters of waxy-white, slightly fragrant blossoms at the tips of its branches. There are only two living species: one in Nevada and western Utah, and the other in a broad but spotty distribution in the Rockies and Southwest. A related species has been found in fossil beds near Creede, Colorado. These date from about 35 million years ago, so *Jamesia* comes from an ancient lineage. Torrey and Gray based their description on material collected by James, but the exact locality in which he collected it is uncertain. In their description, Torrey and Gray state that they "have applied the present name in commemoration of the scientific services of its worthy discoverer, the botanist and historian of 'Major Long's Expedition to the Rocky

Mountains, in the year 1820,' and who, during the journey made an excellent collection of plants under the most unfavorable circumstances."

Before leaving the base camp on Fountain Creek, the naturalists made notes on the burrowing owls so often seen in prairie dog colonies. They captured one and identified it as *Strix* (now *Athene*) *cunicularia*, described some years earlier from Chile by Giovanni Molina. It seemed possible that the birds seen at the base of the Rockies were different, so Say wrote a description, though without providing a name for that population. Burrowing owls occurring in the United States are now considered to belong to the same species as that described from Chile, but to a different subspecies. These are long-legged, mostly diurnal owls that take advantage of prairie dog burrows in which to place their nests. They are not restricted to prairie dog villages, but their numbers have declined in recent years as the "dogs" have been eliminated in many areas.

> JULY 16. JB: At 5 oclock a.m. resumed our march, taking a southardly course, to pass a spur of the mountain on our right, and to arrive on the Arkansas river at the nearest practicable point from our camp, in this, we were obliged to cross deep ravines, over sand hills & ridges, thro thickets of cedar bushes, follow the windings of vallies along the bases of rocky eminences through a dry barron country destitute of water, herbage & game. . . . We experienced more fatigue than at any day we have travelled.

James was even more distressed: "Our sufferings from thirst, heat, and fatigue, were excessive, and were aggravated by the almost unlimited extent of the prospect before us, which promised nothing but a continuation of the same dreary and disgusting scenery." In the afternoon the men reached the Arkansas River, probably about ten miles west of the present city of Pueblo. Here they stopped for two days "to give some of the gentlemen an opportunity of visiting the mountains at the point where the Arkansas river leaves them." Near camp, the hunters killed a doe, "which, though extremely lean, proved an important addition to our supply of provisions." The campsite was an attractive one, shaded by willows and cottonwoods and providing ample pasture for the horses. The men believed that they were near the site of Pike's stockade of 1806, but they were unable to find any evidence of it.

On the morning of July 17, Bell, James, Parish, and Ledoux left camp by horseback, planning to follow the Arkansas to the mountains. Bell reported the going "extremely disagreeable, from the intense heat of the sun [and] the barren broken & knobby surface of the country." So they followed an Indian trail over flatter country to the north, but it led them some eight miles from the river. Heading south to the river, they were rewarded by finding several mineral springs, with unpalatable water and apparently not frequented by herbivorous animals, as saline springs usually are. James called them Bell's Springs. They were located near the site of Canon City, evidently on the grounds of the state penitentiary, but they have since disappeared. "It was near sunset when [the men] arrived at the springs," wrote James, "and being much exhausted by their laborious march, they immediately laid themselves down to rest under the open canopy, deferring their examinations for the following morning."

JULY 18. EJ: In ascending the Arkansa on the ensuing morning, we found the rock to become more inclined, and of a redder colour, as we approached the primitive [rocks], until, at about half a mile from the springs, it is succeeded by the almost perpendicular gneiss rock, which appears here at the base of the first range of mountains. . . . The river pours with great impetuosity and violence through a deep and narrow fissure in the gneiss rock, which rises so abruptly on both sides to such a height, as to oppose an impassable barrier to all further progress.

They had arrived at the foot of Royal Gorge, now a major tourist attraction. Although Pike had explored the upper reaches of the Arkansas much farther (and in winter!), Long had allowed only two days to examine the river as it left the mountains. Once again he had elected not to penetrate the mountains to follow a major river to its sources. Bell expressed disappointment, describing the canyon of the Arkansas as "the grandest & most romantic scenery I ever beheld—what a field is here for the naturalist, the mineralogist, chemist, geologist and landscape painter. I am confident our party has omitted to visit the most interesting spot, where subjects for each department of science is to be found, that will be met with on our whole tour."

But the men were forced to return to the main camp, this time along the river, which in places was bounded by precipices on each side. Seven miles from the mountains, they passed "a remarkable mass of sandstone rocks, resembling a huge pile of architectural ruins [evidently a butte near the present town of Florence]." From this point, the peak that James had climbed was due north. When they reached the main camp, they were tired and hungry. They had traveled sixty-six miles in two days.

James had nothing to say concerning any plants he may have encountered on his trip to the mountains; doubtless there was little time for botanizing. Say was, however, busy at camp. He studied and described two of the most characteristic rodents of the Rocky Mountains. Neither is likely to have been taken at the expedition's campsite; more likely, Bell's party had brought specimens from the foothills. The first of these, a "very handsome species," Say named *Sciurus quadrivittatus* (now *Tamias quadrivittatus*, Colorado chipmunk): "Its nest is composed of a most extraordinary quantity of the burrs of the xanthium [cocklebur] branches, and other portions of the large upright cactus, small branches of pine-trees and other vegetable productions, sufficient in some instances to fill the body of an ordinary cart." Say was surely mistaken about the nest. The nest he described is clearly that of a wood rat, or "pack rat" (*Neotoma*).

Titian Peale, watercolor of Colorado chipmunks. (American Philosophical Society)

The second species had reddish coloration on the head and a pair of broad, lateral, pale stripes on the body. Say named it *Sciurus* (now *Spermophilus*) *lateralis*. This is the golden-mantled ground squirrel, which Lewis and Clark had described informally some years earlier. It is one of the most abundant and attractive rodents in rocky areas throughout much of the mountainous West. In parks, nowadays, these squirrel often become quite tame, and will sometimes take food from one's fingers. Like other ground squirrels, they spend the winter months sleeping deep within their burrows. It is interesting that the naturalists failed to find the Rockies' most characteristic tree squirrel, the Abert's, or tassel-eared, squirrel. These elegant squirrels escaped detection until 1851, when Samuel Woodhouse brought back specimens from New Mexico.

Say also described from this locality "a very beautiful species of emberiza," similar to the indigo bunting but having a white and red breast. He named it *Emberiza* (now *Passerina*) *amoena*. This is the lazuli bunting, so called for the brilliant blue colors of the head and back. There were many cliff swallows near the camp, their mud nests plastered to rocky ledges. The nests, Say noted, somewhat resembled "a chymist's retort," with an entrance near the top that "projects a little and turns downward." Peale sketched the birds, and Say described the species, calling it *Hirundo lunifrons*. He had no way of knowing that three years earlier the cliff swallow had been described in its winter home in Paraguay by a French ornithologist, who had called it H. *pyrrhonota*.

Just as James had a notable genus of plants named for him, Say's contributions to ornithology were recognized by the naming of a phoebe taken on the Arkansas River near the base of the mountains. This is Say's phoebe (*Sayornis saya*), one of the West's most attractive flycatchers. The genus *Sayornis* also includes the eastern phoebe and the black phoebe. It was Charles Lucien Bonaparte, a nephew of Napoleon, who honored Say in this way in 1825. Bonaparte had moved to the United States and settled in Philadelphia, where he compiled the first checklist of American birds and expanded Alexander Wilson's *American Ornithology*, with help from Thomas Say. There are also well over 100 insect species named *sayi* by various entomologists, just as there are many western plants named *jamesii*.

"A fine species of serpent was brought into the camp by one of the men," Say reported. "It moves with great rapidity, and in general form and size it resembles C. *constrictor* [the racer]." Say named it *Coluber testaceus* (now *Masticophus flagellum testaceus*, western coachwhip). This is a slender, fast-moving snake that sometimes climbs into shrubs in its search for prey. A rattlesnake that was captured in a colony of prairie dogs Say described as *Crotalus confluentus*. This was the prairie rattlesnake, which Constantine Rafinesque had named C. *viridis* only two years earlier.

Of course, Say was also collecting insects. Perusal of his later publications reveals at least ten new species "taken on the Arkansas near the mountains." Two of these were large, black beetles that have since become "collectors' items." One was a rather formidable, inch-long tiger beetle that he called *Amblycheila cylindriformis*. The second was a peculiar, stout-bodied longhorn beetle that is usually found associated with cacti: *Moneilema annulata*.

JULY 19. EJ: This morning we turned our backs upon the mountains, and began to move down the Arkansas. It was not without a feeling of something like regret, that we found our long contemplated visit to these grand and interesting objects, was now at an end. One thousand miles of dreary and monotonous plain lay between us, and the enjoyments and indulgences of civilized countries. This we were to traverse in the heat of summer, but the scarcity of game about the mountains rendered our immediate departure necessary.

$\mathcal{N}ine$
DOWN THE ARKANSAS

AS THE EXPEDITION DREW AWAY FROM the mountains, the country became increasingly arid, with, according to James, "scarce a green or a living thing except here and there a tuft of grass, an orbicular lizard, basking in the scorching sand . . . a blaps [darkling beetle], or a galeodes [sun spider]. . . . Near the river, and in spots of uncommon fertility, the unicorn plant . . . was growing in considerable perfection." This remarkable plant, the only member of its family in North America, bears large, tubular blossoms that are replaced by woody pods that terminate in a long, curved appendage suggesting an elephant's trunk (hence the scientific name, *Proboscidea louisianica* [literally, plant with a trunk that inhabits Louisiana Territory]). When the pod dries, the "trunk" splits lengthwise and forms a pair of sharp, upcurved prongs that hook onto the feet of mammals, which carry them off for a distance and in the process disseminate the seeds. Other names for the plant are proboscis-flower, elephant-tusk, ram's-horn, cuckold's horns, double-claw, and devil's claw.

"A large and beautiful animal of the lizard kind . . . was noticed in this day's ride," wrote James. "Its movements were so extremely rapid that it was with much difficulty we were able to capture a few of them." Say described the species as *Ameiva tesselata*. This is the checkered race runner (*Cnemidophorus tesselatus*). It is an unusually

Titian Peale, sketch of a checkered race runner, July 10, 1820. (From the sketch-books of Titian Ramsay Peale, Yale University Art Gallery, gift of Ramsay MacMullen, M.A.H. 1967)

handsome lizard with a complex, checkerboard pattern on its back. It is now known that almost all individuals are females and that females kept in isolation lay eggs that produce fertile female offspring. It is believed that the species may have originally resulted from hybridization between two other species, but it is maintained in nature by having largely dispensed with the male sex and thus the possibility of cross-breeding with other species.

At eleven in the morning of July 19, the men stopped at the mouth of Fountain Creek, entering from the north, "to refresh" and to send out hunters. In the afternoon, they passed the mouth of the St. Charles River, flowing in from the south. The campsite that evening was on a "grassy point on the north side of the river," where they dined on venison and turkey while the mosquitoes dined on them. They had traveled twenty-five miles that day, "over a dusty plain of sand and gravel, barren as the deserts of Arabia," wrote Bell (this was his second evocation of the Arabian deserts, the first being on the South Platte not far from the mountains). But Bell was consoled by the thought that they were traveling toward home: "The anticipation of again enjoying the benefits & pleasure of civilized society and the fond welcome of our friends, cheers our hearts &

gives full scope to fancied imagination in anticipated pleasures perhaps never to be realized."

August Chouteau and Jules de Munn had rendezvoused on a tributary of the St. Charles River in 1817, and Zebulon Pike had passed the river still earlier, in 1806. All three, and their men and provisions, had ended up in the hands of the Spanish. Ezekiel Williams had trapped the upper Arkansas in 1811, and then traveled down the river, leaving his partner, Jean Baptiste Champlain, to be killed by the Arapahos. So Long and his men were retracing the steps of several other groups of Americans, but knowledge of the fate of these earlier treks in the Arkansas Valley cannot have been reassuring.

One of the hunters had encountered a grizzly bear, which approached him; "without staying to make inquiry into the intentions of the animal, [he] mounted his horse and fled." These bears were well known to trappers and traders, and Lewis and Clark as well as Pike knew them well, but they had been formally described only a few years earlier by Philadelphia zoologist George Ord, who named the species *Ursus horribilis*. James's *Account* includes a description of the bear drawn up from a museum specimen after the expedition had returned, along with a series of anecdotes gathered from various sources concerning the size and ferocity of these animals. No narrative of the West of those days would be complete without a few bear stories!

On July 20, the men moved another twenty-six miles, passing the mouth of a tributary from the south that James called "Wharf creek, probably from the circumstances of its washing the base of numerous perpendicular precipices." He had misunderstood the Spanish word *huerfano* (orphan), which the creek was called because farther south the stream passes near an isolated butte of volcanic rock.

Mosquitoes continued to plague the campsites at night, and game was scarce aside from a few turkeys, two deer, and a wildcat. The naturalists took a flycatcher with a yellow breast and a small orange spot on top of the head. Say provided a description, naming the bird *Tyrannus verticalis*, a name that still stands for the western kingbird. It seems odd that they had not encountered this species sooner, as it is a common and widely distributed western bird.

JULY 21. EJ: We left our encampment at five A.M., and having descended six or eight miles along the river, met an Indian and squaw, who were, as they informed us, of the tribe called Kaskaia; by the French Bad-hearts. [Kaskaias are sometimes called Prairie Apaches.] They were on horseback; and the squaw led a third horse of uncommon beauty. They were on their way from the Arkansa below to the mountains near the sources of the Platte, where their nation sometimes resides. They informed us that the greater part of six nations of Indians were encamped about nineteen days' journey below us, on the Arkansa. . . . These nations . . . had been for some time . . . engaged in a warlike expedition against the Spaniards. They had recently met a party of Spaniards on the Red river, when a battle was fought, in which the Spaniards were defeated with considerable loss.

We now understood the reason of a fact which had appeared a little remarkable; namely, that we should have traveled so great an extent of Indian country as we had done since leaving the Pawnees, without meeting a single savage. The bands . . . had all been absent from their usual haunts on a predatory excursion. . . .

The two Indians remained with the expedition for a few miles and were asked about a good place to ford the river and to find a route to the Red River. There was an exchange of gifts, the Indians supplying some jerked bison meat in return for tobacco, a mirror, and other trinkets. Bell also traded a mule, an old dragoon jacket, and other objects for the horse being led by the squaw.

EJ: The Indian informed us he was called "The Calf." He appeared excessively fond of his squaw; and their caresses and endearments they were at no pains to conceal. It was conjectured by our guide, and afterwards ascertained by those who descended the Arkansa, that they had married contrary to the laws and usages of their tribe, the woman being already the wife of another man, and run away for concealment.

After covering only fourteen miles, the party camped on the river among cottonwoods and willows. The understory consisted of false indigo (*Amorpha fruticosa*), three species of milkweeds (*Asclepias*), two species of sunflowers (*Helianthus*), "the great bartonica" (presumably giant evening star, *Mentzelia decapetala*), prickly poppy (*Argemone*), and the usual cacti. There were also sedges and grasses (wild rye and bluestem). The hunters brought in "two deer, one antelope, and seven turkeys."

The expedition was now about 100 miles from the mountains, near the site of the present-day town of Rocky Ford. Long elected to remain there for two days, making preparations to divide the party, with one group to continue down the Arkansas, and the other to search toward the south for the Red River.

Two kinds of bats were found at the campsite, one of them previously unknown. Say described it, naming it *Vespertilio subulatus*; this is now called *Myotis subulatus*, or the small-footed myotis. This is a widely distributed, yellowish-brown bat that roosts in small groups in caves, trees, and buildings.

It was probably at this camp that Say took three rather formidable wasps, all three illustrated in color in his *American Entomology*. The largest of them had a blue-green body about an inch and a half long and bright orange wings spanning nearly three inches. Say named it *Pompilus formosus* (now *Pepsis formosa*): "This large and splendid species occurred within a hundred miles of the Rocky Mountains, on the banks of the Arkansas river. It was not uncommon, and in consequence of the striking color of the wings, as well

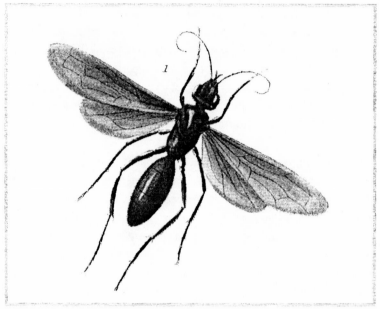

Titian Peale, drawing of a tarantula hawk. Described by Thomas Say from the upper Arkansas Valley, this wasp was drawn for Say's *American Entomology.*

as of its slow and steady flight, it was readily observed and taken." Species of this genus are known to prey on tarantulas and place them in shallow burrows in the soil, where the wasps' eggs are laid and their larvae develop at the expense of the paralyzed tarantula. They are often called "tarantula hawks." Say evidently did not have the experience of being stung by one of these monsters, or he surely would have made a note of it.

Another wasp, nearly as large but banded with yellow and having transparent wings, he named *Stizus* (now *Sphecius*) *grandis.* Say surmised that because of its resemblance to the cicada killer wasp of the East, this species probably also provisioned its nest with cicadas. In this he was later proved correct. Along with these wasps he took a smaller species, black with a single orange band on the abdomen, that he called *Stizus* (now *Stizoides*) *renicinctus*. This wasp, we now know, is a brood parasite of other ground-nesting wasps. All three of these wasps remain locally common along the Arkansas Valley.

On July 24, Long, James, Peale, and seven others forded the river and headed south in an effort to find the Red River and so accomplish one of the goals of the expedition. This ford was well known to the Indians and was later used by traders on their way to Santa Fe. Today Rocky Ford is best known for its cantaloupes.

The remainder of the party would follow the Arkansas downstream to Fort Smith, which Long had helped to establish in 1817. The group included Bell (who was appointed leader), Swift, Say, Seymour, and six others (Julien, Parish, Nowland, Barnard, Foster, and Myers). Bijeau and Ledoux would accompany the group part of the way, and then leave to return to their homes in the Pawnee villages on the Loup River of Nebraska. The two dogs that had accompanied the expedition all the way from Engineer Cantonment would join Bell's contingent. The two groups would meet at Fort Smith (on the western border of present-day Arkansas).

After Long and his companions had crossed the river, they "gave three cheers & took off in a southardly course over the prairie," as Bell put it. I shall take up the story of Long's detachment in Chapter 10.

JB: The Commanding officer selected the ablest and best horses and mules for the party destined to the Red River, believing that party would have a greater distance to travel and over more broken country; otherwise, an impartial division of stock on hand was made, here follows an Invoice of what was furnished [Bell's] detachment—viz: 14 horses, 2 mules ... many of them worn out with fatigue & sore back, that could not be used. Indian goods, for presents to the Indians and to purchase from them necessary articles, viz:—6 doz. knives, 31 looking glasses, 1 tin case, 26 small twists tobacco, 1½ lb vermillion, 5½ doz. combs, 115 small bells, 15 fire steels, 6 pair scissors, 3 bunches beads, 9 moccoson awls, 4 gun worms and 86 flints—!! what a pitiful stock for an exploring party fitted out by the government of the United States, what a contemptable opinion must the savages form of our nation, it is too bad. Of provisions we have 9 pints of corn, 3 pecks parched corn meal, 12 lb biscuit, ¾ lb coffee, ½ lb sugar, 1 oz tea, 5 pints whiskey, 4 bottles Lemn Acid—and of ammunition—2 lb powder & 31 small cans of lead!!! Horse equipments and miscellaneous articles—10 saddles, 2 bridles, 3 bear skins, 5 provision bags, 1 ax & sling, 2 common tents, 1 flag or colour, 1 melting ladle, packing lines, fishing lines & hooks, a number of packages containing collections made by the party, a small tin case containing assorted medicine, 3 camp kettles,—Our powder horns, recently filled with powder, and we have a small quantity of jirked meat cured by the party yesterday. Thus have I stated, as correctly as possible the outfit of our detachment, destined to find our way through bands of wandering Savages, war parties, and over a wilderness country, 800 miles to white settlements.

At ten o'clock on July 24, the much diminished exploring party proceeded down the north side of the Arkansas River, camping that night not far from the site where, in 1833, William Bent would build a fort that became a major stopover on the Santa Fe Trail. There was now no botanist, so there are few remarks on the vegetation they found along their way. This section of the expedition's report, as stated in the *Account*, is "from the pen of Mr. Say." Bell, of course, continued his journal, an additional source of information.

On July 25, the party passed the mouth of a large tributary from the south, the Purgatoire. This river had been named by the Spanish "Rio de las Animas Perdidas en Purgatorio" (River of Lost Souls in Purgatory) because of several deaths that had occurred there without redemption from a priest. French traders had shortened the name to

Purgatoire, and to early American settlers, it was the Picketwire. The city of Las Animas now stands near the confluence.

> JULY 26. TS: Late in the afternoon we saw, at a great distance before us, evident indications of the proximity of Indians, consisting of [tipis], on the edge of the skirting timber, partially concealed by the foliage of the trees. On our nearer approach we observed their horses grazing peacefully, but becoming suddenly frightened, probably by our scent, they all bounded off towards the camp, which was now in full view. Our attention was called off from the horses by the appearance of their masters, who were now seen running towards us with all their swiftness. A minute afterwards we were surrounded by them, and were happy to observe in their features and gestures a manifestation of the most pacific disposition; they shook us by the hand, assured us by signs that they rejoiced to see us, and invited us to partake of their hospitality. We however replied, that we had brought our own lodges with us, and would encamp near them. We selected for this purpose a clear spot of ground on the bank of the river, intending to remain a day or two with this little known people, to observe their manners and way of life.

The Indian camp contained members of four nations: Kiowas, Kaskaias, Cheyennes, and Arapahos. On the following day, chiefs of each of the tribes met with the explorers, who attempted to impress on them that they were not traders but were studying the landscape and the plants and animals. Indeed, they had little to trade, but did their best to satisfy the Indians, while obtaining food and even horses in exchange. Seymour painted the camp of the explorers, showing their tents surrounded by Kiowas. He also made portraits of some of the chiefs.

They were able to communicate with the Indians in a roundabout way. Bell spoke to Julien in English, and Julien translated into

Engraving in the *Account*, after Samuel Seymour, watercolor of Kaskaia, Cheyenne, and Arapaho chiefs.

French for Ledoux. Ledoux knew Pawnee and transmitted the messages to an Indian who knew that language, and this person translated them into the tongues of the local chiefs. In spite of these problems, Bell apparently made a fairly long speech, explaining what the explorers were doing in Indian lands and promising that the Great White Father would "render every assistance" to his "Red children" if they would be peaceful and "live like brothers." The chiefs replied that they were happy to see the Americans and that their "war parties should be instructed to shake hands & make [friends] with the Americans where ever they met them."

JB: The first evening of our arrival among them, the Chiefs politely offered me the use of one of their wives, during the time we should remain among them—which I as politely declined, they said we had no squaws with us and must necessarily want them & it was with some difficulty I could object to their solicitations without offending them—during each night we remained with them several of the Indians brought their wives into camp & remained all night, the husband going around to the members of our party soliciting as a favour, connection with his wife, for a small piece of tobacco or a little vermillion— during the time, the wife would be laying on a buffalo robe covered with another, they were not generally their youngest or handsomest wives. . . .

TS: Soon after our arrival, an Indian well stricken in years inquired if we had seen a man and squaw within a day or two on our route: we described to him the appearance of the calf and his squaw. "That is my wife," he said, "who has eloped from me, and I will instantly go in pursuit of them." He accordingly procured a companion, and both were soon on their way, well armed and mounted.

Say described the appearance and clothing of the Indians in some detail. He was particularly anxious to record their languages, which "abound with sounds strange to our ears." He did not fail to make note of their intimacy with certain insects.

TS: In the rear of our tent, a squaw, who had become possessed of a wooden small-toothed comb, was occupied in removing from her head a population as numerous, as the individuals composing it were robust and well fed. She had placed a skin upon her lap to receive the victims as they fell; and a female companion who sat at her feet alternately craunched the oily vermin between her teeth, and conversed with the most rapid and pleasant loquacity, as she picked them up from the skin before her.

W. W. Wood, drawing of crested-keel grasshoppers. Say discovered this striking species in the Arkansas Valley and illustrated it in his *American Entomology.*

On July 28, the Indians departed upstream, while the exploring party moved downstream through a violent thunderstorm. "A fine species of toad (bufo) inhabits this region," reported Say. He described it as *Bufo cognatus,* now known as the Great Plains toad. These toads, we now know, range throughout much of the Great Plains and Southwest, breeding in ponds and in areas that are temporarily flooded after rains. According to Geoffrey Hammerson, in *Amphibians and Reptiles in Colorado,* in the breeding season the male produces "an ear-splitting trill which sounds almost like a jackhammer."

It may have been on this date that Say collected two unusually large and ornate grasshoppers, both of which were illustrated on a color plate in his *American Entomology.* These were *Gryllus* (now *Tropidolophus*) *formosus* and *G.* (now *Acrolophitus*) *hirtipes.* Both have a high crest on the thorax and are elaborately patterned with green, with conspicuously banded wings.

Like many an entomologist, Say admired tiger beetles and doubtless enjoyed collecting them. These attractively patterned beetles run rapidly over the ground in pursuit of smaller insects, which they grasp in their large jaws. They take flight readily when pursued by an entomologist. Along the Arkansas, Say found and described a relatively large, "remarkably splendid" species with a reddish-coppery sheen to the wing covers. He called it *Cicindela pulchra.* A still larger but mainly black species that was "not uncommon on the banks of the Arkansa river" he called C. *obsoleta.* Individuals of this species provide a special challenge to collectors; when disturbed, they fly for many feet, and then drop quickly to the ground and blend with the soil.

As the party proceeded along the river, they had several further meetings with Indians, all of them more or less friendly. Some of the Indians carried cakes of pemmican made of choke cherries, stones and all, mixed with bison fat. Bell declared them "quite delicious." Say was compiling a vocabulary of Indian words and recording what he could of their customs. On one occasion, he induced an Arapaho medicine man to open the contents of his bag.

JULY 30. TS: At our solicitation he readily opened his sacred depository, and displayed its contents on a skin before us, whilst he politely proceeded to expatiate on their powers and virtues in the occult art, as well as their physical efficacy. They consisted of various roots, seeds, pappus, and powders, both active and inert, as respects their action on the human system, carefully enveloped in skins, leaves, &c., some of which, to his credulous faith, were invested with supernatural powers. Similar qualities were also attributed to some animal products with which these were accompanied, such as claws of birds, beaks, feathers, and hair. But the object that more particularly attracted our attention was the intoxicating bean, as it has been called, of which he possessed upwards of a pint. Julien recognized it immediately. He informed us, that it is in such high request amongst the Oto Indians, that a horse has been exchanged for eight or ten of them. . . . With some few trinkets of little value, we purchased the principal portion of our medicine man's store of beans; they are of an ovate form, and of a light red, sometimes yellowish colour, with a rather deeply impressed oval cicatrix, and larger than a common bean. . . .

The beans were presumably those of mesquite (*Prosopis*), which when ground up and allowed to ferment in water produce a beerlike beverage. As the party moved down the river, Bell remarked that the plants he saw "differ almost every day. It is a matter of regret that not one of our detachment is sufficiently skilled in botany to note and describe the vegetable productions."

. On about August 1, the expedition crossed what is now the Colorado–Kansas state line. There was a scare when a war party of Cheyennes, painted and in full battle dress, galloped toward them, shouting and brandishing their bows and arrows. Bell ordered their flag unfurled and their horses staked and guarded by soldiers with loaded rifles. But the Cheyennes had come from a skirmish with the Pawnees and were afraid of being pursued by them, so they moved on after the usual pipe-smoking and exchange of gifts, the chief leaving Bell with a hug.

Horseflies were abundant along the river bottoms, often covering the necks of the horses and "dyeing them with blood." The horses were already "sufficiently miserable," and there is no doubt that loss of blood was one reason for their increasingly weakened condition. Rattlesnakes were also abundant, but the horses seemed to avoid them successfully, and neither horses nor men were bitten. Burrowing owls were seen, and it was clear that the burrows they occupied were in poor condition, obviously those of prairie dogs that had been abandoned by their owners.

AUGUST 5. TS: [T]he lowing of the thousands of bisons that surrounded us in every direction, reached us in one continual roar. This harsh and guttural noise, intermediate between the bellowing of the domestic bull and the grunting of the hog, was varied by the shrill bark and scream of the jackals [coyotes], and the howling of the . . . wolves. . . . These wild and dissonant sounds were associated with the idea of the barren and inhospitable wastes, in the midst of which we were then reposing, and vividly reminded us of our remoteness from the comforts of civilized society.

"Not having had a supply of fresh buffalo meat for nearly a month," Bell commented, "we this day feasted on it, of the choicest pieces, boiled and roasted & made into what the interpreters call hunters pudding." Say noted among the herds of bison flocks of "cow bunt-ings," which he called *Emberiza pecora*. This is a name formerly used for the brown-headed cowbird (*Molothrus ater*). Cowbirds once may have been restricted to following grazing mammals of the West, where they acted as nest parasites of birds of open country. Over time, as forests were cleared for the grazing of cattle and the growing of grains, cowbirds spread over much of the continent, where they came to parasitize many songbirds. Nowadays, they are blamed, in part, for the decline of many songbird species throughout North America.

On August 6, the explorers found the river turning north. As they realized, they had reached the beginning of the "great bend" of the Arkansas (they were a few miles east of the site of the present city of Dodge City, Kansas). Bijeau and Ledoux now left the party and headed for their homes on the Platte.

TS: We cannot take leave of them, without expressing our entire approbation of their conduct and deportment during our arduous journey; Bijeau, particularly, was faithful, active, in-dustrious, and communicative. Besides the duties of guide and interpreter, he occasionally and frequently volunteered his services as hunter, butcher, cook, veterinarian, &c., and pointed out various little services, tending to our comfort and security, which he performed with pleasure and alacrity, and which no other than one habituated to this mode of life would have devised. During leisure intervals, he had communicated an historical narrative of his life and adventures, more particularly in as far as they were relative to the country which we have been exploring. . . .

A copious vocabulary of words of the Pawnee language was obtained from Ledoux, together with an account of the manners and habits of that nation. . . .

As the party, diminished by two, proceeded along the river, Say spoke of the "debilitating influence" of the extreme heat—over ninety degrees virtually every day. Their horses were now becoming very weak; there was little feed for them, as the bison had eaten most of the grass. On August 8, the men passed the mouth of "Vulture creek, from the number of that bird seen about it." This was probably the Pawnee River, which reaches the Arkansas near the modern-day town of Larned, Kansas. Sunflowers were everywhere, and along the river there were not only cottonwoods, but also elms, walnuts, mulberries, and ashes, "which we hail with a hearty welcome, as the harbingers of a more productive territory," wrote Say.

On August 10, they reached the apex of the bend, not far from the present-day city of Great Bend. From that point, the river flows in a generally southeasterly direction. The hunters brought in a "fine fat deer" and the meat of a cow bison. "We have truly feasted to day," wrote Bell, "our dessert after dinner consisting of sour grapes and black walnuts." Say reported a bald eagle "sailing high in the air," and the "rice bird (emberiza oryzivorus L.) . . . feeding on the seeds of sunflower." This was the bobolink, now called *Dolichonyx oryzivorus*.

Two days later, Say was "gratified with the appearance of the prairie fowl [lesser prairie-chicken] running nimbly before us through the grass, the first we have seen since leaving the Platte." Later that day, there was an episode with a party of Comanches, but once again the expedition passed through without serious trouble. On the following day, Swift shot a large elk, which was cut up and added to the party's dwindling food supply. It was Sunday, but they did not stop to rest, as they feared meeting with other war parties and, said Bell, "their friendship is only to be purchased by tobacco, of which we are out."

AUGUST 14. TS: [O]ur morning's journey was arduous, in consequence of the great heat of the atmosphere. Our dogs, these two or three days past, had evidently followed us with difficulty. Caesar, a fine mastiff, and the larger of the two, this morning trotted heavily forwards and threw himself down directly before the first horse in the line; the rider turned his horse aside, to avoid doing injury to the dog. . . . The dog, finding this attempt to draw attention to his sufferings unavailing, threw himself successively before two or three other horses, but still failed to excite the attention he solicited, until a soldier in the rear observed that his respiration was excessively laborious, and his tongue to a great length depended from his widely extended mouth. He therefore took the dog upon his horse before him, intending to bathe him in the river, [but] the poor exhausted animal expired in his arms before he reached it. To travellers in such a country, any domesticated animal, however abject, becomes an acceptable companion; and our dogs, besides their real usefulness as guards at night, drew our attention in various ways during the day, and became gradually so endeared to us, that the loss of Caesar was felt as a real evil.

That day the party passed the mouth of the Little Arkansas, at the site of the city of Wichita. There were honey locusts and sycamores as well as cottonwoods, elms, and ashes. Quail and prairie chickens were common. On the following day, the men came on a deserted Indian village, where they took some corn and melons. On August 14, Say had recorded a prairie dog village, which proved to be the last they saw as they plodded southeastward. Two days later, it was noted that the antelope had disappeared, and bison and their accompanying wolves were rarely seen. Oaks, walnuts, willows, and mulberries filled a ravine. Say found puffball mushrooms (*Lycoper-don*) "nearly equal in size to a man's head." These puffballs are ed-ible—indeed, quite delicious when properly prepared—but Say and his men may have been unaware or uncertain of this. The men also found a "plant familiarly known in the settlements by the name of Poke, (Phytolacca *decandra*.) [now *P. americana*, often called poke-weed]." This is a lush green plant with deep purple berries, the sole North American representative of a family of mostly tropical distri-bution. The expedition had now crossed the ninety-eighth meridian, often considered the approximate dividing line between the semiarid West and the more humid East.

The day's march on August 16 brought the expedition past the mouth of the Ninnescah River, where there were great numbers of turtles. Game was now scarce, but the next day the hunters were able to take three turkeys and two fawns. Common elders (*Sambucus*) were seen. The following day they passed Walnut Creek and were approximately at the present state line between Kansas and Oklahoma. Hickories were seen for the first time since the expedi-tion had left the Missouri River.

On August 19, the second dog, Buck, died, even though one of the men had carried him on horseback because of his weakened condition. The men were tired of the meanderings of the river and the numerous gulleys that they had to cross. "If we had but a map of the river," Bell remarked, "how many bends & hills & hollows & ravines we could have avoided!!" The next day the party dined on "a few mouldy biscuit crumbs, boiled in a large quantity of water, with the nutritious addition of some grease." Late in the day, Julien killed a skunk, which was added to their soup of crumbs the following day (it "tasted skunkish enough," Bell commented). On August 22, the hunters took nothing and were forced to rescue the body of a small fawn from the wolves that had killed and partially devoured it. The scientific expedition had turned into a race for survival. But Say took occasion to note the presence of a spectacular bird.

AUGUST 22. TS: A note like that of the prairie dog for a moment induced the belief that a village of the marmot was near; but we were soon undeceived by the appearance of the beautiful tyrannus forficatus [scissor-tailed flycatcher] in full pursuit of a crow. Not at first recognizing the bird, the fine elongated tail plumes, occasionally diverging in a furcate manner, and again closing together, to give direction to the aerial evolutions of the bird, seemed like the extraneous processes of dried grass, or twigs of a tree, adventitiously attached to the tail, and influenced by currents of wind. The feathered warrior flew toward a tree, from whence, at our too near approach, he descended to the earth at a little distance, continuing at intervals his chirping note.

Say reported on August 23, that the explorers were "once again saluted by the note of the blue jay." A flock of Carolina parakeets flew by, and kingfishers and warblers were sighted. "A large white crane (ardea egretta of Wilson) stalked with slow and measured strides in the shallows of the creek," Say reported. This was presumably a great egret (*Casmerodius albus*). Say also captured a glass snake, which he called "ophisaurus ventralis" (now *Ophisaurus attenuatus*, eastern glass snake). Glass snakes are not true snakes, but legless lizards.

AUGUST 25. TS: Remained encamped in order to give the hunters an opportunity to procure some game. We had nothing for breakfast or dinner, and as our meals a few days past had been few and slight, we have become impatient under the pressure of hunger; a few fresh-water muscles (unio), two or three small fishes, and a tortoise which had been found in the mud of the ravine, were roasted and eaten, without essential condiment salt, of which we had been for some time destitute. The hunters so anxiously looked for at length returned, bringing but three ducks. . . .

As we have no idea of our distance from Belle Point [where Fort Smith was located], and know not what extent of country we are doomed to traverse in the state of privation to which we have of late been subjected, we have selected, from our miserable horses, an individual to be slaughtered for food, in case of extremity of abstinence; and upon which, although very lean, we cannot forbear to cast an occasional wishful glance.

The following day the men found some green plums and ate as many as they "thought safe." Bell noted that a little "Milnor's Lemon acid," which they were carrying, made the water more palatable and helped relieve their hunger a bit. The hunters no longer had shoes or moccasins. "Poor fellows," wrote Bell, "they have to hunt thro' briers bare foot." But Swift managed to kill a large buck deer, whereupon, said Bell, "a smile of joy lighted every countenace & every jaw wagged in anticipation."

The Arkansas Valley was becoming so thick with brush that a few days later they elected to follow an Indian trail away from the river. But it took them through hilly country where the going was slow and game still scarce. There were small flocks of "the common wild pigeon," presumably mourning doves. Say's horse, "a sprightly, handsome, and serviceable animal" that had accompanied him since Nebraska, was now completely exhausted and had to be abandoned. Parish's horse was also unable to rise and was left behind. The party returned to the river late on August 31, but only after an experience that caused their morale to plummet still further.

AUGUST 31. TS: We arose early, and on looking at the horses that were staked around the camp, three of the best were missing. Supposing that they had strayed to a distance, inquiry was made of the corporal respecting them; who answered that three of the men were absent, probably in pursuit of them, and added, that one of those men who chanced to be last on guard had neglected to awaken him to perform his duty on the morning watch. Forster [Foster], a faithful, industrious soldier . . . now exclaimed, that his knapsack had been robbed; and upon examining our baggage, we were mortified to perceive that it also had been overhauled and plundered during the night. But we were utterly astounded to find that our saddle bags, which contained our clothing, Indian presents, and manuscripts, had also been carried off.

This greatest of all privations that could have occurred within the range of possibility, suspended for a time every exertion, and seemed to fill the measure of our trials, difficulties, and dangers.

It was too obvious that the infamous absentees, Nolan [Nowland], Myers, and Bernard [Barnard], had deserted during the night, robbing us of our best horses, and of our most important treasures. We endeavoured in vain to trace them, as a heavy dew had fallen since their departure, and rested upon every spear of grass alike, and we returned from the fruitless search to number over our losses with a feeling of disconsolateness verging on despair.

Our entire wardrobe, with the sole exception of the rude clothing on our persons, and our entire private stock of Indian presents, were included in the saddle bags. But their most important contents were all the manuscripts of Mr. Say and Lieut. Swift, completed during the extensive journey from Engineer Cantonment to this place. Those of the former consisted of five books, viz. one book of observations on the manners and habits of the Mountain Indians, and their history, so far as it could be obtained from the interpreters; one book of notes on the manners and habits of animals, and descriptions of species; one book containing a journal; two books containing vocabularies of the languages of the Mountain Indians; and those of the latter consisted of a topographical journal of the same portion of our expedition. All these, being utterly useless to the wretches who now possessed them, were probably thrown away upon the ocean of prairie, and consequently the labour of months was consigned to oblivion by these uneducated vandals.

The three deserters, Say added, had proved "worthless, indolent, and pusillanimous from the beginning," and Nowland was known to have deserted on two previous occasions. To a degree, the desertion of the men from the floundering and half-starved expedition can be excused. They had no way of knowing that very soon conditions would improve and that within nine days the party would reach their destination at Fort Smith.

JB: With heavy hearts & sad countenances we arranged what was left into packs as light as possible, disposing of all that was deemed unnecessary or superfluous, in order to relieve our wearied horses. . . . Halted at mid-day to refresh, not being able to travel in open prairie country more than at the rate of 2½ miles an hour, altitude of the mercury 94 degrees.

SEPTEMBER 1. JB: All our party was out at day light in quest of game, returned to Camp about 8 oclock without having killed anything, brought in a few sour grapes which satisfied our hunger a little. We do not suffer much the pangs of hunger, but keep constantly growing weaker & losing flesh, when off our horses, feel no disposition to walk, but to sit or lay down, exercise produces a trembling and pain in the knees & legs. Before 9 oclock we proceeded along the bottom where the travelling for some time was easy & pleasant, when it became interupted by ravines & low scrubby bushes. In these, Lieut. Swift killed a young deer, & Julian collected a quantity of plums on the margin of the river—when we arrived at a convenient place we halted, on the margin of the bank, turned out our horses to pasture, and ourselves to cooking some of the deer, on which and the plums, we feasted most sumptuously, being the first meal we had eaten in four days. It is impossible to describe the immediate restorative effects, this meal produced upon our feelings and strength. Thus revived we resumed our way at 3 p.m. weather extremely warm. . . .

Soon an Osage Indian approached, the first of a hunting party that arrived on the following day. "This noble generous Indian," wrote Bell, "finding we had been nearly starved, gave us from his robe a quantity of plums he had collected and what tobacco he had for us to chew & smoke which was the greatest luxury of all. . . . [H]e took one of our rifles and went out to kill a deer for us."

When the main band arrived, they prepared a feast for the explorers. The Osages had had considerable contact with whites, and many of their implements had been acquired by trade with settlements in Missouri. Their chief, Iron Bird—or Clermont, as he was known to the whites—had been visited by Lieutenant James Wilkinson, under orders from Zebulon Pike, in 1807, and Thomas Nuttall had met with him on his trek up the Arkansas River in 1819. He had even been to Washington. In 1834, his portrait was painted by George Catlin. Clermont received Bell hospitably, but was unable to supply him with the horses and guides that Bell had hoped for.

The Indians reported that three white men had been seen in their village. Swift, Julien, and some of the Indians set forth after them, since they evidently were the deserters. However, they returned in the evening after discovering that the three had already left the village. There seem to be no records of what became of Nowland, Myers, and Barnard. Since they were army privates, they would have been punished severely if in fact they reached settlements and were recognized and reported. The valuable notebooks of Say and Swift were doubtless "thrown away upon the ocean of prairie," as Say surmised. It is remarkable that Say did as well as he did in reconstructing his itinerary and his notes and descriptions of animals.

The meeting with the Osages must have occurred near the site of the thriving city of Tulsa, Oklahoma. On the afternoon of September 3, Bell and his men left the Indians and crossed a prairie with "pleasing groves of oak" as well as pecans (*Carya illinoensis*), which were seen for the first time. Hunting was improving, and along with gifts from the Indians they were fairly well off.

SEPTEMBER 4. JB: We are now making up for our long spell of starvation, we have abundance of venison, corn, pumpkin & squash, we eat our three meals a day in such quantities at a time as would astonish any person who had not witnessed the recovery of persons after having [been] in a state of starvation . . . distance travelled to day 17 ¾ miles.

The expedition had now passed the mouth of the Verdigris River, flowing from the north. This part of the Arkansas Valley had been followed by Say's friend Thomas Nuttall just a year earlier. By this time Nuttall was so weakened by starvation and fever that he was often delirious and at times had to be helped by a companion to mount his horse. Nevertheless he brought back a notable collection of plants and insects. James and Say must surely have read Nuttall's *Journal*, which was published in 1821, when they prepared their own *Account* in 1823. However, they make no mention of the fact that they were now traversing country already explored by a well-qualified naturalist.

On September 5, the men reached a trading post where they were "hospitably received" and served a meal at a table supplied with stools and benches, luxuries they had not experienced in some time. "Our pleasure at first meeting civilized white men was of no ordinary kind," wrote Say. At the trading post, Say took "a beautiful species of lizard . . . [that] runs with great swiftness." He provided a description, naming it *Agama collaris*. This is the eastern collared lizard, now called *Crotaphytus collaris*. It is indeed a beautiful species, yellowish with dark bands, blue spots, and a black "collar."

On leaving the trading post, the party crossed through a dense

canebrake and saw sassafras trees and flowering dogwoods, further
evidence that they were now in a region of higher rainfall. The next
day brought them to "Mr. Bean's salt works," where they were
treated to a drink of buttermilk that Bell found "more gratifying to
our palates, than the sparkling champain."

SEPTEMBER 6. TS: Whilst waiting with a moderate share
of patience for our evening meal of boiled pumpkins, one of
the children brought us a huge hairy spider, which he carried
upon a twig, that he had induced the animal to grasp with its
feet. Its magnitude and formidable appearance surprised us. The
boy informed us that he had captured it near the entrance of
its burrow, and that the species is by no means rare in this part
of the country. Not having any box suitable to contain it, nor
any pin sufficiently large to impale it, we substituted a wooden
peg, by which it was attached to the inside of a hat. This species
so closely resembles, both in form, colour, and magnitude, the
gigantic bird-catching spider of South America, that from a
minute survey of this specimen, which is a female, we cannot
discover the slightest characteristic distinction.

This was, of course, a tarantula (more properly, a theraphosid spider),
similar to but not identical to the giant tropical spider mentioned
by Say. Smaller but more insidious relatives of spiders were also making themselves evident: ticks. Say found them crawling "by dozens
up our leggings" and causing "an intolerable itching." Evenings were
spent picking "the pestiferous arachnides" from their bodies. There
were two kinds: the larger was the size of a head of a pin; the smaller,
barely visible. Say included a description of the larger kind, naming
it *Ixodes molestus*. Probably the "two kinds" represented young and
adults of the lone star tick (*Amblyomma americanum*), which is still
a serious pest in that area. Linnaeus had described the species more
than half a century earlier.

SEPTEMBER 8. TS: On a naked part of the soil, gullied out by the action of torrents of water, we beheld a hymenopterous or wasp-like insect ... triumphantly, but laboriously, dragging the body of the gigantic spider, its prey, to furnish food to its future progeny. We cannot but admire the prowess of this comparatively pigmy victor, and the wonderful influence of a maternal emotion, which thus impels it to a hazardous encounter, for the sake of a posterity which it can never know.

The wasp was clearly a spider wasp (Pompilidae), and if by the words "the gigantic spider" Say was referring to the kind he had reported two days earlier, then the wasp was a tarantula hawk (*Pepsis*), similar to the species he had described from the Arkansas Valley not far from the Rockies.

On September 9, the exhausted band finally reached Fort Smith, where they were greeted by Captain James Ballard, who was in charge in the temporary absence of Major William Bradford. "His politeness and attention," wrote Say, "soon rendered our situation comfortable, after a houseless exposure in the wilderness of ninety-three days."

Bell calculated that they had traveled 873¼ miles since the first camp on the Arkansas. He felt that "the satisfaction of our safe arrival was almost destroyed by the absence of the deserters with important manuscripts. . . . And in addition, where was the commanding officer & his party? they should have arrived before this time, if there has been no cause of detention; so that altogether, our feelings on our arrival was made of a combination of indescribeables."

While Bell's contingent waited for the arrival of Long and his group, they were well fed and reclothed, and worked on their reports and maps. One further effort was made to trace the deserters, with the help of the Osage Indians, and a reward of $200 was offered for their capture. But these efforts were in vain.

Fort Smith was situated on a hill overlooking the junction of the Arkansas and Poteau Rivers. It was surrounded by forests of oaks, tulip trees, and sassafras, with an understory of grapes, smilax, and other vines that rendered it almost impassable except by trails that had been cut. Settlers in the area suffered various illnesses, including "bilious fevers," sometimes aggravated by "the destructive habits of intemperance." But there were gardens in which corn, melons, sweet potatoes, "and other esculent vegetables" were grown. To the exploring party, Fort Smith must have seemed like paradise.

$\mathcal{T}en$

THE SEARCH FOR THE RED RIVER

MAJOR LONG, WITH JAMES, PEALE, AND seven others (Adams, Dougherty, Wilson, Duncan, Oakley, Verplank, and Sweney), set off from the Arkansas River not far from the site of Rocky Ford, Colorado, on July 24. They had six horses and eight mules, mostly in reasonably good condition. Heading south, in 100-degree heat, they traveled twenty-seven miles that first day, finding barely enough water to supply their needs and nothing but bison dung for fuel. James found a new and attractive coneflower, which he described as *Rudbeckia* (now *Ratibida*) *tagetes*. He also noted yellow flax (*Linum rigidum*) and a species of globe mallow. The latter was later described by John Torrey as *Sida stellata* (now *Sphaeralcea angustifolia cuspidata*).

The next day they crossed several ravines, some of which contained box elders (*Acer negundo*). Near midday, they struck the Purgatoire River, in a valley flanked by sandstone cliffs nearly 200 feet high. They followed the Purgatoire for only a few miles before entering "the valley of a small creek, tributary from the south-east." This was undoubtedly Chacuaco Creek, which they followed for two days. (The expedition's route is by no means easy to follow over the next ten days. In "Major Long's Route from the Arkansas to the Canadian River, 1820," John M. Tucker has worked out the itinerary in convincing detail, and I shall follow his interpretation.)

Titian Peale, watercolor of a line-tailed squirrel (rock squirrel). (American Philosophical Society).

It was rough going, with masses of fallen rocks, thickets of alders and willows, and muddy places in the streambed. James noted "the yellow-bellied fly catcher [likely a western kingbird] and the obscure wren [rock wren]." A different species of ground squirrel was taken on the sandstone cliffs, and later described by Say as *Sciurus grammurus* (now *Spermophilus variegatus grammurus*, rock squirrel). Examination of the mouth pouches showed them to be filled with the buds and leaves of plants growing among the rocks. The rock squirrel has a bushy but somewhat flattened tail, edged with white. Say called it the "line-tailed squirrel." Nearly two centuries later, little is known about the life style of these squirrels, though they are known to range sparingly throughout much of the West.

JULY 26. EJ: A beautiful dalea [prairie clover], two or three euphorbias [spurges], with several species of eriogonum [wild buckwheats], were among the plants collected about this encampment. Notwithstanding the barrenness of the soil and the

aspect of desolation which so widely prevails, we are often surprised by the occurrence of splendid and interesting productions springing up under our feet, in situations that seemed to
promise nothing but the most cheerless and unvaried sterility.
Operating with unbounded energy in every situation, adapting
itself with wonderful versatility to all combinations of circumstances; the principle of life extends its dominion over these
portions of nature which seem as if designed for the perpetual
abode of inorganic desolation, distributing some of its choicest
gifts to the most ungenial regions; fitting them by peculiarity
of structure, for the maintenance of life and vigour, in situations
apparently the most unfavoured.

At nine o'clock in the evening of the 25th, a fall of rain
commenced; we were now ten in company, with a single tent,
large enough to cover half the number. In order, however, to
make the most equal distribution of our several comforts, it was
so arranged that about the half of each man was sheltered under
the tent, while the remainder was exposed to the weather. This
was effected by placing all our heads near together in the centre
of the tent, and allowing our feet to project in all directions,
like the radii of a circle.

On the following day, the party left the canyon they were following,
"without the least regret," and emerged on "a boundless and varied
landscape." George Goodman and Cheryl Lawson make the case
that they had now left the Chacuaco and had been following a tributary, Bachicha Canyon. On the plains, "herds of bison, antelopes,
and wild horses gave life and cheerfulness to the scene." A species
of purple-flowered *Gaura* was common; James described it and
named it *G. mollis*. Gauras are members of the evening primrose
family and are sometimes called butterfly weeds, bee blossoms, or
wild honeysuckles. The flowers are borne on tall spikes and bloom
from the lower portion toward the top. The unusual blossoms have
four petals at the top and a group of stamens below. Linnaeus had

perhaps named the genus *Gaura*, based on the Greek word for "proud," because of the curious, uplifted petals. James collected several species of the genus on the expedition.

The hunters wounded a bison, which ran off pursued by wolves. Fortunately, they were able to catch up with it and kill it. Although it was dark by that time, the meat was prepared, and the men "spent the greater part of the night regaling on the choice pieces."

Before reaching their campsite, the men had noted a "naked pile of rocks towering to a great elevation" to the east. This was probably the western promontory of Mesa de Maya. "James Peak" was sighted, but (according to Tucker) it was probably Greenhorn Mountain, north of the present city of Walsenburg. The campsite of July 27 was evidently a few miles east of the present-day town of Branson, Colorado.

On July 28, the men passed a hill of "green stone" (volcanic rock), possibly Negro Mesa, which lies on the present Colorado–New Mexico border. Soon they began a descent between walls of sandstone. This (according to Tucker) was probably Tollgate Canyon, which acquired its name some years later when, for a time, travelers on the Santa Fe Trail were required to pay a fee for passage. A canyon a few miles to the east has come to be called Long's Canyon, but it is probable that Long did not actually follow this canyon. Both lead to the Cimarron, which Long's group reached on the evening of July 28. They had no name for the stream, but believed it to be a tributary of the Canadian; in fact, it joins the Arkansas separately from the Canadian. It was a relief to come upon a stream of clear water, free of alkali.

During the day James had noted magpies, horned larks, and cowbirds. Cholla cacti and wild gourds were common. At their campsite along the stream, the men were serenaded by robins and mockingbirds from the oaks and cottonwoods. James wrote that "the stern features of nature, which we had long contemplated with a feeling almost of terror, seemed to relax into a momentary smile to cheer us on our toilsome journey."

On the following day, they proceeded south through two violent storms accompanied by wind and hail. The rain continued until dark, with the temperature falling to forty-seven degrees. There was no fuel and little food, so all ten men piled into the tent to "restore the warmth" of their "benumbed bodies" by placing their bodies together "in the least possible compass." Peale became ill, but was "somewhat relieved by the free use of opium and whiskey."

JULY 30. EJ: We left our comfortless camp at an early hour on the ensuing morning, and traversing a wide plain . . . we arrived in the middle of the day in the sight of a creek, which, like all the watercourses of this region, is situated at the bottom of a deep and almost inaccessible valley. With the customary difficulty and danger, we at length found our way down to the stream, and encamped.

We were much concerned, but by no means surprised, to discover that our horses were rapidly failing under the severe services they were now made to perform. We had been often compelled to encamp without a sufficiency of grass, and the rocky travelling, to which we had for some time accustomed them, was wearing out and destroying their hoofs. Several were becoming lame, and all much exhausted and weakened. . . .

The stream which may be supposed to exist in [the valley] for a part of the year at least, but which was now dry, runs towards the south-east. Having arrived at that part of the country which has by common consent been represented to contain the sources of the Red river of Louisiana, we were induced, by the general inclination of the surface of the country and the direction of this creek, to consider it as one of those sources; and accordingly resolved to descend along its course, hoping it might soon conduct us to a country abounding in game, and presenting fewer obstacles to our progress that that in which we now were. Our sufferings from the want of provisions, and from the late storm, had given us a little distaste for prolonging farther than was necessary our journey towards the southwest. And our horses failing so rapidly, that we did not now believe they would hold out to bring us to the settlements by the nearest route.

Various opinions have been expressed about which stream Long's group had reached and was to follow for the next five days. Long's map identifies it as the Rio Mora, a river that arises on the eastern slopes of the Sangre de Cristo Mountains and flows mostly south and then east. However, there is every reason to believe that the expedition was many miles to the east of the Rio Mora. Tucker has presented convincing evidence that the stream was Ute Creek, which the party first met near the site of the modern town of Gladstone (Union County), and followed to near the site of Logan (Quay County), both in New Mexico.

Between the Purgatoire and entry into the valley of Ute Creek, James had collected "many new and interesting plants." They included evening star (*Mentzelia*), raspberry (*Rubus*), milkvetch (*Astragalus*), beard-tongue (*Penstemon*), scorpion-grass (*Myosotis*), and sunflower (*Helianthus*). The *Penstemon* is believed to be the one that George Bentham later described as *P. jamesii*. James also noted purslane and "a very small cuscuta" (dodder) that was parasitic on the purslane.

Mule deer had been seen many times since the expedition arrived near the Rockies, and had often provided venison for the men. The naturalists were under the impression that the species had never been formally described, though Lewis and Clark and others who had traveled in the West had seen mule deer many times. So on July 31, the hunters were sent out and a reward was offered if they would bring in an intact, mature male. Verplank took one, but it was not brought in until dark. Since the men needed the meat for supper, Peale quickly took appropriate measurements and made a sketch. The head and hide were kept as specimens, and the rest was devoured. Say later described the species, using this material and the

Titian Peale, watercolor of a mule deer. (American Philosophical Society)

measurements made in the field. He named it *Cervus macrotis* (*macrotis* being Greek for "long-ears"). All of this proved to be in vain, as the mule deer had been described in 1817 by Constantine Rafinesque on the basis of a specimen collected by Lewis and Clark in the Dakotas. He named it *C. hemionus* (*hemionus* being Greek for "mule"). Now the mule deer is called *Odocoileus hemionus*.

Peale's sketch of the deer is extant and includes some of the landscape in the background, including a mesa of distinctive shape. When tracing the route of the expedition along Ute Creek, Tucker found and photographed a mesa of identical appearance, further demonstrating that he had correctly identified the route. Another stream farther east (Tramperos Creek) was often called Major Long's Creek on early maps, on the incorrect assumption that it was the stream that Long and his men had descended.

At their camp along what the explorers believed to be a tributary of the Red River, James collected a sensitive briar (*Schrankia*) and two species of butterfly weed (*Gaura*). Rattlesnakes were seen, as well as two kinds of horned lizards that differed in the length of their spines and the position of their nostrils. Probably these were *Phrynosoma cornutum* and *P. douglassii*. There were old lava flows and other evidences of past volcanic activity.

As they continued down the valley in a southeasterly direction, the explorers endured several storms. At their camp on August 2, they could find neither firewood nor bison dung for fuel, so they could not cook a badger they had shot and had to substitute "the eighth part of a sea biscuit each," which was to supply them with both supper and breakfast. The occasional stagnant pools in the streambed did little to satisfy their thirst or that of their horses. Strains were developing among the travelers. "The weather continues warm," James wrote in his diary, "and we are growing tired of each other, and of our comfortless and weary pilgrimage."

James discovered a small tree that puzzled him. It produced a "leguminous fruit" in the form of seedpod from six to ten inches long, in which the seeds were enclosed in separate cells "immersed in a saccharine pulp" that was "very grateful to the taste when ripe." The leaves were pinnate, and the trunk was covered with simple spines. He had discovered honey mesquite, a tree of wide distribution in the Southwest that John Torrey later named *Prosopis glandulosa* on the basis of samples collected by James. Mesquite was a staple food among many southwestern Indians, who made a nutritious meal from the pods and fermented the fruits to make an intoxicating drink.

AUGUST 3. EJ: [A]s the prospect of the country before us promised no change, it is not surprising we should have felt a degree of anxiety and alarm, which, added to our sufferings from hunger and thirst, made our situation extremely unpleasant. We had travelled great part of the day enveloped in a burning atmosphere, sometimes letting fall upon us the scorching particles of sand, which had been raised by the wind, sometimes almost suffocating ... when we had the good fortune to meet with a pool of stagnant water, which, though muddy and brackish, was not entirely impotable, and afforded us a more welcome treat than it is in the power of abundance to supply. Here was also a little wood, and our badger, with the addition of a young owl, was very hastily cooked and eaten.

Numbers of cow buntings [cowbirds] had been seen a little before we arrived at this encampment, flying so familiarly about the horses that the men killed several with their whips.

As the expedition moved down the valley, the water that appeared in the streambed contained "a quantity of red earth as to give it the colour of florid blood." They took this as a confirmation of their belief that this must be a tributary of the Red River. There were several kinds of cacti, some of them with edible fruits. James collected partridge pea (*Cassia*); false indigo (*Amorpha*); bull nettle or stinging bush (*Jatropha*, now *Cnidoscolus*); and other plants.

Wild horses were seen, and one of them approached closely enough to be shot. "We had all suffered so severely from hunger, and our present want of provisions was so great," wrote James, "that instead of questioning whether we should eat the flesh of a horse, we congratulated ourselves on the acquisition of so seasonable a supply. We felt a little regret at killing so beautiful an animal ... but our scruples all yielded to the loud admonitions of hunger."

Although the *Account* is vague on this point, the party had evidently begun to follow an Indian trail that departed from the stream they had been following. On the evening of August 5, they camped on a river "sixty yards in width, twenty of which were naked sand-bar, the remaining forty covered with water, having an average depth of about ten inches." The water was "intensely red, having nearly the temperature and the saltiness of new milk." To these remarks from the *Account* can be added a comment from James's diary that they now believed themselves to be "on the main Red River and not on one of its branches." According to Tucker, Ute Creek flows in a deep canyon before entering the river, and the expedition had apparently missed the actual junction by following the Indian trail. Long and his men were to follow the "main Red River" for many days.

James continued to botanize, finding a gentian, a *Croton*, a broomrape, a parasitic plant lacking chlorophyll (*Orobanche*), and others: "The common partridge (perdix virginianus) was seen near this encampment, also the dove, which had never disappeared entirely in all the country we had passed." The partridge was presumably the bobwhite (now *Colinus virginianus*); the doves were doubtless mourning doves.

In his diary, James remarked on prairie dogs standing erect beside their burrows: "A scene of this sort comprises most of what is beautiful or interesting in the plain and woodless country which constitutes so great a part of the territory of Louisiana."

Peale killed a burrowing owl and examined the contents of its crop, finding it filled with grasshopper wings and parts of other insects. This removed suspicions that the owls might prey on the prairie dogs whose villages they inhabited. Sandburs (*Cenchrus*) had become very common, their spiny fruits "falling into our mockasins, adhering to our blankets and clothing, and annoying us at every point." Cockleburs (*Xanthium strumarium*) were ripe, "adding one more to the sources of constant molestation."

AUGUST 7. EJ: A formidable centipede (scolopendra) was caught near the camp, and brought in alive by one of the engagees. It was about eight inches in length, and nearly three fourths of an inch in breadth, being of a flattened form, and of a dark brown colour. While kept alive, it showed great viciousness of disposition, biting at every thing which came within its reach. Its bite is said to be venomous.

Titian Peale, ink sketch of a burrowing owl, August 7, 1820. (American Philosophical Society)

There were tracks of bison, giving hope that the men might soon see "the return of the days of plenty." The expedition crossed several tributaries of the river it was following. Most were dry, but the size of their valleys suggested that they must drain a wide expanse of arid country subject only to occasional summer cloudbursts and spring snowmelt. They had now crossed the present border between New Mexico and the panhandle of Texas.

These were lean days indeed. The hunters spent much time "in an unavailing search after game." On August 9, the men ate the last of the horse they had shot four days earlier; "the weather since [having been] unusually warm, [it] had suffered from long keeping." Again the hunters had no success. James expressed the party's concern: "Our suffering from want of provisions, and from the apprehension of still more distressing extremities, were now so great, that we gave little attention to any thing except hunting." Although they had no way of knowing it, Bell's contingent on the Arkansas was having very similar experiences!

Major Long expressed his opinion of the country they were traversing in his report to Secretary of War John Calhoun, submitted soon after his return to the East in January 1821.

> SL: In regard to this extensive section of the country, I do not hesitate in giving the opinion, that it is almost wholly unfit for cultivation, and of course uninhabitable by a people depending upon agriculture for their subsistence. Although tracts of fertile land considerably extensive are occasionally to be met with, yet the scarcity of wood and water, almost uniformly prevalent, will prove an insuperable obstacle in the way of settling the country. This objection rests not only with the section immediately under consideration, but applies with equal propriety to a much larger portion of the country. . . . This region, however, viewed as a frontier, may prove of infinite importance to the United States, inasmuch as it is calculated to serve as a barrier to prevent too great an extension of our population westward. . . .

For these remarks and the map he prepared, in which the southern High Plains are labeled "Great Desert," Long has often been severely criticized. But he was not alone in these feelings. Spanish nobleman Francisco Vásquez de Coronado, in 1541, remarked that "it was the Lord's pleasure that, after having journeyed across the deserts seventy-seven days, I arrived at . . . Quivira." (Quivira proved to be in central Kansas.) Zebulon Pike, after crossing the southern plains in 1806, opined that "these vast plains . . . may become in time equally celebrated as the sandy deserts of Africa." Thomas Nuttall, exploring the lower Arkansas Valley for plants in 1819, spoke of the "inhospitality of this pathless desert." Edwin James wrote that any "traveller who shall at any time have traversed its desolate sands, will, we think, join us in the wish that this region may for ever remain the unmolested haunt of the native hunter, the bison and the jackall." John Bell twice spoke of parts of the country they had crossed as "barren as the deserts of Arabia," and Thomas Say wrote to a friend that the region within 500 miles of the Rockies was "totally unfit for the tillage of civilized man."

Two decades later, Josiah Gregg, who knew the Santa Fe Trail from personal experience and wrote about it in *Commerce of the Prairies*, spoke of crossing "this dreaded desert." Crossing Nebraska on the Oregon Trail in 1846, Francis Parkman spoke of the "barren, trackless waste, extending for hundreds of miles to the Arkansas on the one side, and the Missouri on the other. . . . Sometimes [the plain] glared in the sun, an expanse of hot, bare sand; sometimes it was veiled by long, coarse grass." In 1873, John Hanson Beadle, in his book *The Undeveloped West*, assured his readers that "the Great American Desert [is] not a myth."

These men (aside from Coronado) were accustomed to the well-watered and forested East; they can have had little conception of the profound effects of limited rainfall on the landscape. Even today, Easterners who cross the western plains are impressed (if not appalled) by the apparent bleakness of the countryside—especially in those parts through which the Long Expedition was now passing.

These impressions must also be evaluated in the context of their times. In the early nineteenth century, the country was indeed "pathless," and wood and water were in short supply nearly everywhere. That the landscape would someday be criss-crossed by paved highways, the surface waters dammed and ditched, and the groundwater tapped could not have been foreseen by explorers too intent on mere survival to dream of technologies yet to be invented.

The word "desert" was and is appropriate. In a cool climate, any region that receives an average of under ten inches of precipitation annually is defined as a desert; in a hot climate, with greater evaporation, twenty inches is the usual figure given. The High Plains have fiercely hot summers (as Long's party was finding out), but winters can be bitterly cold. So if we assume that fifteen inches is the maximum precipitation for such a climate to provide desert conditions, the region is a desert or, at best, a semidesert. Studies of tree rings have shown that 1820 was a year of severe drought throughout the Southwest, so Long was experiencing landscapes even more arid than usual. Throughout history, the West has been subject to alternating periods of drought and more ample rainfall.

Even today, some of the country the Long Expedition traversed during August 1820 is sparsely settled and marginally suitable even for cattle. But some of the country the explorers passed through, particularly the river valleys and the land along the Front Range, is now lush with food crops, and cities have sprung up in places where Long and his men once camped in the wilds with little food, fuel, or water. Clearly the "desert" did not prove to be a barrier to westward migration. And yet—now that all the rivers have multiple dams and the great aquifers are being seriously depleted—a barrier to further settlement may soon reappear. Rain does not follow the plow, as many who moved west believed, and there are ultimate limits to the amount of water that can be squeezed from the land by even the most sophisticated technologies.

But to return to our story. On August 10, there was prospect of a change in the fortunes of Long's half-starved group, for better or worse, when a party of about 250 Indians approached them. The chief, Red Mouse, shook hands all around and told the explorers that he and his people were Kaskaias on their way to trade with the Spanish. James described Red Mouse as "of large stature . . . somewhat past the middle age of life, and no way deficient in his person, and countenance of those indications of strength, cunning, and ferocity, which form so important a part of greatness in the estimation of the Indians." Red Mouse assured Long that, indeed, the expedition had been traveling along the Red River. The explorers were invited to camp with the Indians, and they did so, not wishing to antagonize the Indians, from whom they hoped to obtain food and horses. They watched the Indians put up their tipis, using poles their horses had dragged from a great distance, as there was no suitable wood for many miles around. Peale set about to sketch their tipis.

This tribe of Indians occupied the southern High Plains, but

Engraving in the *Account*, after Titian Peale, watercolor of skin tipis of the Kaskaias.

often hunted along the Red River and the Brazos. "The great numbers of images of the alligator, which they wear either as ornaments or as amulets for the cure or prevention of disease and misfortune, afford sufficient proof of their extending their rambles to districts inhabited by that reptile," wrote James. "These images are of carved wood covered with leather, and profusely ornamented with beads." Red Mouse had been wounded in the arm by an arrow some time previously, and he used an ornamented image of an alligator, pressed repeatedly in his hand, as a "cure" for his wound.

James reported that "the men are expert horsemen, and evince great dexterity in throwing the rope, taking in this way many of the wild horses which inhabit some parts of their country. They hunt the bison on horseback with the bow and arrow, being little acquainted with the use of fire arms." There was evidence that the

Titian Peale, pencil sketch of an Indian bison hunt. (American Philosophical Society).

Indians had had some contact with the Spanish, but this may have been their first contact with whites of English descent. They took pleasure in admiring the white skin of the arms of the explorers.

Although the young women were "far from disgusting in their appearance," as they matured they became "squabbish." James added, "Their breasts become so flaccid and pendulous that we have seen them give suck to their children, the mother and child at the same time standing erect upon the ground." Their usual garment was a loose frock, but it was often discarded in favor of a small leather apron. As with other tribes, their bodies were home to many lice, which the women ate "with avidity."

The Indians were disappointed with the few items that Long and his men had to trade, and at one point tried to open their packs looking for more, resulting in a scuffle. Long's men were better armed than the Indians, who soon backed off. But there was little hope of obtaining much from the Kaskaias, whose hospitality extended only to serving the men "a little half-boiled bison meat," from which the squaws had taken the best pieces for their children. They offered the men water from the "paunch of a bison" that still retained its original smell. The next morning, the Indians prepared to move, after absconding with several of the explorers' horses, kettles, and other equipment. Long ordered his men to seize some of the Indians' possessions, and after a show of force persuaded the Indians to return most of what they had taken. "We parted . . . as friends," wrote James. But it had been a stressful meeting, and left the explorers with a low opinion of the Kaskaias.

> EJ: Though we saw much to admire among this people, we cannot but think they are among some of the most degraded and miserable of the uncivilized Indians on this side of the Rocky Mountains. Their wandering and precarious manner of life, as well as the inhospitable character of the country they inhabit, precludes the possibility of advancement from the profoundest barbarism. As is common among other of the western tribes, they were persevering in offering us their women, but this appeared to be done from mere beastliness and the hope of reward, rather than from any motive of hospitality or a desire to show us respect. We saw among them no article of food except the flesh of the bison; their horses, their arms, lodges, and dogs, are their only wealth.

On August 12, Long and his men "moved on rather briskly," not wishing to take a chance that some of the Indians might make a further effort to steal their horses. At their camp along the river that evening, they saw avocets, terns, and other water birds. There were incrustations of salt resulting from the evaporation of seepages along the bank. James noted the presence of various plants "delighting in a saline soil," including saltbush, winged pigweed, and goosefoot.

The temperature remained well over 100 degrees as the party moved on downstream. On one occasion, Peale became separated from the others and spent the night alone, harassed by mosquitoes as he slept under some trees where bison had recently been; the place offered "little refreshment," he reported. On August 14, James saw blue jays, purple martins, and turkeys. Plants characteristic of more humid areas were appearing: elms, pokeweed (*Phytolacca*), and buttonbush (*Cephalanthus*).

AUGUST 15. EJ: Several species of locust [cicada] were extremely frequent here, filling the air by day with their shrill and deafening cries, and feeding with their bodies great numbers of that beautiful species of hawk, the falco Mississipiensis of Wilson [now *Ictinia mississippiensis*, Mississippi kite]. It afforded us a constant amusement to watch the motions of this greedy devourer in the pursuit of his favourite prey, the locust [cicada]. The insect being large, and not uncommonly active, is easily taken; the hawk then pauses on the wing, suspending himself in the air, while, with his talons and beak, he tears in pieces and devours his prey.

We were fortunate in capturing a tortoise. . . . The upper part of its shell was large enough to contain near a quart of water, and was taken to supply the place of one of our tin cups recently lost, while the animal itself was committed to the mess kettle. Wolves, jackals [coyotes], and vultures, occurred in unusual numbers, and carcasses of several bison recently killed had been seen. . . . Near our camp was a scattering grove of small-leaved elms. This tree (the U. alata, N.) is not known in the Eastern states.

According to Goodman and Lawson, the trees noted by James were probably small American elms (*Ulmus americana*). Two weeks later, James again noted *U. alata* (winged elm or wahoo), but by that time the travelers were well within the known range of that species.

The river was paralleled by sand dunes, and blowing sand was a major annoyance to both men and horses. On August 16, there was a violent thunderstorm, with hailstones measuring nearly an inch in diameter. Bison were converging in great numbers to drink from stagnant pools in the riverbed. The men found it better to make shallow wells in the sand to obtain drinking water.

AUGUST 17. EJ: The small elms along this valley were bending under the weight of innumerable grape vines, now loaded with ripe fruit, the purple clusters crowded in such pro- fusion as almost to give a colouring to the landscape. On the opposite side of the river was a range of low sand hills, fringed with vines, rising not more than a foot or eighteen inches from the surface. On examination, we found these hillocks had been produced exclusively by the agency of the grape vines, arresting the sand as it was borne along by the wind, until such quantities had been accumulated as to bury every part of the plant, except the end of the branches. Many of these were so loaded with fruit, as to present nothing to the eye but a series of clusters, so closely arranged as to conceal every part of the stem. The fruit of these vines is incomparably finer than that of any other na- tive or exotic which we have met with in the United States. The burying of the greater part of the trunk, with its larger branches, produces the effect of pruning, inasmuch as it pre- vents the unfolding of the leaves and flowers on the parts below the surface, while the protruded ends of the branches enjoy an increased degree of light and heat from the reflection of the sand. . . . We indulged ourselves to excess, if excess could be committed in the use of such delicious and salutary fruit, and invited by the cleanness of the sand, and a refreshing shade, we threw ourselves down, and slept away, with unusual zest, a few of the hours of a summer afternoon.

Our hunters had been as successful as could be wished, and at evening we assembled around a full feast of "marrow- bones," a treat whose value must for ever remain unknown to those who have not tried the adventurous life of the hunter. We were often surprised to witness in ourselves a proof of the facility with which a part at least of the habits of the savage could be adopted. . . .

Clearly there were days when the lives of the explorers were not filled with the stresses produced by heat, thirst, hunger, and the uncertainties of tomorrow. Besides the grapes, wild plums were also abundant. There was evidence, from animal droppings, that the plums and grapes were being fed on not only by turkeys and black bears, but even by such carnivores as wolves and coyotes. Black walnuts now appeared along the stream, as well as "the great flowering hibiscus [rose mallow]," which was a "highly ornamental plant among the scattering trees in the low grounds." The men had now crossed the hundredth meridian, which forms the boundary between the Texas panhandle and Oklahoma.

By August 18, Long was becoming concerned because the river seemed to be flowing northeastward, which "did not coincide entirely with our previous ideas of the direction of the Red river." However, the assurance of the Indians they had met with a week earlier "tended to quiet the suspicions we began to feel on this subject."

AUGUST 18. EJ: At sunset we pitched our tent on the north side of the river, and dug a well in the sand, which afforded a sufficient supply of wholesome, though brackish, water. Throughout the night the roaring of immense herds of bisons, and the solemn notes of the hooting owl were heard, intermixed with the desolate cries of the [coyote] and the screech-owl. The mulberry, and the guilandina [Kentucky coffee tree], growing near our camp, with many of the plants and birds we had been accustomed to see in the frontier settlements of the United States, reminded us of the comforts of home and the cheering scenes of civilized society, giving us at the same time the assurance that we were about to arrive at the point where we should take leave of the desert.

AUGUST 19. EJ: Notwithstanding the astonishing numbers of bison, deer, antelopes, and other animals, the country is less strewed with bones than almost any we have seen; affording an evidence that it is not a favourite hunting ground of any tribe of Indians. The animals also appear wholly unaccustomed to the sight of man. The bisons and wolves move slowly off to the right and left, leaving a lane for the party to pass, but those on the windward side often linger for a long time, almost within the reach of our rifles, regarding us with little appearance of alarm. We had now nothing to suffer either from the apprehension or reality of hunger, and could have been content that the distance between ourselves and the settlements should have been much greater than we supposed it to be.

The expedition was now in attractive country, whose extensive grasslands promised to be as good for grazing cattle as they were for supporting bison. In rocky crevices grew penstemons and evening primroses. But nature was not always benevolent. Blowflies were so abundant that even as the men set about to consume their meat, their "table often became white with the eggs deposited by these flies." They found it expedient to make a soup and leave the meat "immersed in the kettle until we were ready to transfer it to our mouths." Gnats and ticks had also become abundant. Black bears were now seen frequently, feeding on grapes, plums, the berries of red osier and round-leaved dogwood, and the "acorns of a small scubby oak, common about the sand hills." The men found the flesh of the bears "deserving of the high encomiums lavished upon it."

On August 21, there was a series of thunderstorms, and at night the thunder was "so blended with the roaring of the bisons, that more experienced ears than ours might have found a difficulty in distinguishing between them." On the following morning, the men found that so much rain had fallen during the night that "they enjoyed the novel and pleasing sight of a running stream of water." It had been two weeks, and over 150 miles of travel, since they had had running water in the stream they were following. Various tributaries, like the main stream, had been "beds of naked sand."

AUGUST 24. EJ: Our supply of parched corn meal was now entirely exhausted. Since separating from our companions on the Arkansa, we had confined ourselves to the fifth part of a pint each per day, and the discontinuance of this small allowance was at first sensibly felt. We however became gradually accustomed to the hunter's life in its utmost simplicity, eating our bison or bear meat without salt or condiments of any kind, and substituting turkey or venison, both of which we had in the greatest plenty, for bread. The few hungry weeks we had spent about the sources of the river had taught us how to dispense with superfluous luxuries, so the demands of nature could be satisfied.

The inconvenience we felt from another source was more serious. All our clothing had become so dirty as to be offensive both to sight and smell. Uniting in our persons the professions of traveller, hostler, butcher, and cook, sleeping on the ground by night, and being almost incessantly on the march by day; it is not to be supposed we could give as much attention to personal neatness as might be wished. . . .

The common post oak, the white oak, and several other species, with gymnocladus or coffee-bean tree, the cercis [redbud] and the black walnut, indicate here a soil of very considerable fertility; and game is so abundant, that we have it at any time in our power to kill as many bison, bear, deer, and turkies as we may wish, and it is not without some difficulty we can restrain the hunters from destroying more than sufficient to supply our wants. Our game today has been two bears, three deers, one turkey, a large white wolf, and a hare. . . .

All of this in what is now central Oklahoma, not many miles west of Oklahoma City! There were also many water birds, suggesting that there were bodies of water nearby. There were killdeers, sand-pipers, yellow-shanked snipes, and telltale godwits (the last two are early names for the lesser and greater yellowlegs, respectively). Other birds included cardinals, summer tanagers, scissor-tailed flycatchers, and pileated woodpeckers. Crows had replaced ravens. "Thickets of oak, elm, and nyssa [tupelo], began to occur on the hills," wrote James, "and the fertile soil of the low plains to be covered with a dense growth of ambrosia [ragweed], helianthus [sunflower], and other heavy weeds." The alternation of forests and luxuriant mead-ows, with plenty of game, led the men to feel that "the habitation and the works of man alone seem wanting to complete the picture of rural abundance."

Many of the plants they were now encountering were those they were familiar with from the eastern states. They included car-dinal flower (*Lobelia cardinalis*), balloon vine (*Cardiospermum hali-cacabum*), and copperleaf (*Acalypha*). Trees included black locust, Ohio buckeye, and persimmon. Mistletoe grew from the branches of the elms. There were no longer any cacti, yuccas, prickly poppies, or other plants of the country close to the mountains.

AUGUST 27. EJ: We found . . . the annoyance of innu-merable multitudes of minute, almost invisible wood ticks, a sufficient counterpart to the advantages of our situation. These insects, unlike the mosquitoes, gnats, and sand flies, are not to be turned aside by a gust of wind or an atmosphere surcharged with smoke, nor does the closest dress of leather afford any protection from their persecutions. The traveller no sooner sets foot among them, then they commence in countless thousands their silent and unseen march; ascending along the feet and legs, they insinuate themselves into every article of dress, and fasten, unperceived, their fangs upon every part of the body. . . . If the insect is suffered to remain unmolested, he protracts his feast for some weeks, when he has found to have grown of enormous size. . . . It is not on men alone that these blood-thirsty insects fasten themselves. Horses, dogs, and many wild animals are subject to their attacks. On the necks of horses they are observed to attain a very large size.

Very likely these were ticks of the same species that Say and his companions were encountering on the Arkansas. Long's party passed many carcasses of bison that had recently been slaughtered, presum-ably by a passing group of Indians. "[T]he air was darkened by flights of carrion birds," both turkey and black vultures. Trees included black cherry, linden, and honey locust, "all affording indications of a fertile soil." There was a colony of honeybees near the camp, as-suring the party that they were close to civilization. "Bees, it is said by the hunters and the Indians," wrote James, "are rarely if ever seen more than two hundred and fifty or three hundred miles in advance of the white settlements." Some of the men went to the bee tree and brought back some honey "enclosed in the skin of a deer recently killed."

On August 29, Adams, the Spanish interpreter, strayed from the party while looking for his canteen and became thoroughly lost. He followed the river for five days, without being able to take any game, before he was found by Long and his men sitting before a fire, starving and in "the deepest despondency."

Traveling in the streambed was difficult, but the woods pro-vided even more difficult going, since they were thick with vines, including greenbriar (*Smilax*) and Virginia creeper (*Parthenocissus*). On September 1, Carolina parakeets appeared, and there were syc-amores like those along the Ohio and Mississippi Rivers. From a hilltop, there was a view toward the north of "a wild and mountain-ous region, covered with forests, where, among the brighter verdure of the oak, the nyssa [tupelo], and the castanea pumila [chinquapin], we distinguished the darker shade of the juniper, and others of the coniferae." James was understandably excited to see the day-to-day change from the flora of the semiarid western regions to that of the lush eastern forests.

The next day the explorers found the remains of several garfish, their thick skins resisting decay and the attacks of scavengers. James listed the trees he now saw along the valley: hackberry, cottonwood, pecan, ash, mulberry, and others. There were squirrels in the trees. A large flock of white pelicans passed on their way up the river. About the camp were snowy egrets and "great numbers of cranes, ducks, pelicans, and other aquatic birds." A black wolf was seen, and the men attempted without success to take it, thinking that it might be a different species from the gray, or "white," wolf. "An enormous black hairy spider" was seen, doubtless a tarantula similar to the one that Say had seen on the Arkansas.

September 5 found the party in fertile country, filled with interesting plants that unfortunately were mostly past flowering. They saw the first osage orange trees (*Maclura pomifera*). (The genus had been named by Thomas Nuttall for his friend William Maclure, who had accompanied Say and Peale on their earlier trip to Florida.) The orangelike fruit of this tree had been reported to be edible when ripe, but James found it "disagreeable to the taste." The juice, when applied to the skin, seemed to afford some protection from ticks.

A bird call was heard that sounded like "the noise of a child's toy trumpet." It proved to be that of an ivory-billed woodpecker, then fairly widely distributed in the South, though now extinct. Pileated woodpeckers were also seen, along with parakeets, mockingbirds, chuck-will's-widows, bald eagles, vultures of two species, and snowy and great egrets.

On September 6, the men passed a sandstone ridge that crossed the river, producing a waterfall. James believed it to be the only spot within 600 miles that could be identified by description.

> EJ: In ascending, when the traveller arrives at this point, he has little to expect beyond, but sandy wastes and thirsty inhospitable steppes. . . . Beyond, commences the wide sandy

desert stretching westward to the base of the Rocky Mountains. We have little apprehension of giving too unfavorable an account of this portion of the country. Though the soil is, in some places, fertile, the want of timber, of navigable streams, and of water for the necessities of life, render it an unfit residence for any but a nomad population.

Several miles below the falls, a large tributary came in from the north. The amount of water in the river was now so great that it was difficult to proceed in the riverbed, especially as there were patches of quicksand. However, the valley was narrow and its sides were covered with trees and vines, so the men preferred to continue along the stream when possible. Beaches and islands in the river were covered with young willows and cottonwoods.

On September 8, the travelers were lucky enough to find an abandoned log canoe, and they loaded it with some of their baggage in order to relieve their tired horses. Two men were assigned to navigate the canoe downstream, but they had to spend much of their time wading and dragging the canoe over the shallow bottom. The next day some twenty or thirty elk were seen along the river, and one was harvested to supplement their diet.

Long had regularly calculated the party's latitude and longitude by making appropriate celestial measurements. It was now clear that they were much too far north to be on the Red River, which they must have suspected for some time. On September 10, when they arrived on the Arkansas, the explorers realized that they had been following the Canadian River, which in fact had sometimes been called the "red river" by the Indians because of the color of the water in its upper reaches. It is probable that the unusual name Canadian for this river was based on the Spanish word *cañada*, with reference to the steep-walled canyons that the river has carved in New Mexico.

It is easy to find excuses for Long's inability to find the true Red River, but in fact he had failed to fulfill still another objective of his expedition. However, his was the first exploration of the Canadian, so the effort was by no means in vain. The large tributary that the men had seen a few miles below the falls was the North Fork of that river.

This was the third attempt the government had made to locate the sources of the Red River, the supposed southern boundary of Louisiana Territory. Pike had tried and failed in 1806, and during the same year Richard Sparks and Thomas Freeman had tried to ascend the river from the east. Both expeditions had been thwarted by the Spanish, and it was not until 1852 that the sources of the Red River were discovered, by Captain Randolph Marcy. The river originates in the high plains of eastern New Mexico and the Texas panhandle. At one point, near Amarillo, Texas, the Red River is only about fifty miles south of the Canadian. At various places in Oklahoma, tributaries of the two rivers are only a few miles apart. But eventually the Red, after forming the boundary between Oklahoma and Texas for many miles, turns southeast, finally to enter the Mississippi deep in the state of Louisiana. The Canadian, though, is a tributary of the Arkansas, as Long now appreciated to his dismay.

Filled with "disappointment and chagrin," Long and his men crossed the Arkansas, only to find themselves surrounded by "a dense and almost impenetrable cane brake, where no vestige of a path could be found." Forcing their way through the tall canes, the party "might be said to resemble a company of rats traversing a sturdy field of grass." They were so battered and exhausted that evening that they slept soundly despite the irritation of ticks—until (not having pitched their tent) they were drenched by a heavy shower in the middle of the night.

Extensive stands of giant cane (*Arundinaria gigantea*) along this section of the Arkansas Valley had been noted by Thomas Nuttall in 1819, and Bell's group had traversed part of the same canebrakes several days earlier. Most of these canebreaks have now been cleared for growing crops and grazing cattle.

The next day the men emerged into open woods, where there

was a "deep morass, covered with the nelumbo [American lotus, or pond nut] and other aquatic plants." Eventually they located the trail to Fort Smith, which Bell and his companions had followed five days earlier. Here they met a group of Indians, who told them that Fort Smith was only a day's ride away. There were pawpaw trees, "with ripe fruit of an uncommon size and delicious flavour." Fallen fruits were "eagerly sought after by the bears, raccoons, oppossum, &c."

At the camp that night were several interesting plants, including Virginia meadow beauty (*Rhexia virginica*), hawkweed (*Hieracium gronovii*), cupseed (*Calycocarpum lyoni*), and butterfly pea (reported as *Vexillaria virginica*). In a footnote, James explained that he had adopted the name *Vexillaria*, from Amos Eaton's *Manual of Botany*, in preference to an earlier and therefore more correct name that had been "severely censured." Linnaeus had named the genus *Clitoria*, since he believed that the keel in the blossoms resembled a mammalian clitoris. *Vexillaria*, based on a Latin word for "flag," James felt to be "equally appropriate." Evidently, it was less threatening to nineteenth-century sensibilities.

In the evening the men received a visit from a trader on his way from Fort Smith to his trading post on the Verdigris River: "the first white man not of our own party whom we have seen since the 6th of the preceding June." The trader gave them coffee, biscuits, and "spirits." He informed them that Bell and his group had arrived at Fort Smith a few days earlier.

On September 13, they "emerged from the deep silence and twilight gloom of the forest" and found themselves "once more surrounded by the works of man." At a farm they passed, they found their "uncouth appearance a matter of astonishment to both dogs and men." Arriving at a beach opposite Fort Smith, they discharged a pistol and were soon ferried across to meet Major William Bradford, Captain James Ballard, and the members of Bell's contingent. Bradford treated them to a meal and cautioned them not to indulge too lavishly on foods that they had not tasted for some time. "The experience of a few days taught us that it would have been fortunate for us if we had given more implicit heed to his caution," wrote James.

The expedition members were once again assembled in one place. Since leaving Engineer Cantonment on the Missouri, they had traveled approximately 1,600 miles by foot and horseback. Of the twenty-two men who had started, seventeen returned; three had deserted, and two had left by prior arrangement. They had lost a few horses and their two dogs, but despite frequent complaints about their condition, all the men had survived in reasonably good health (in contrast to several other exploring parties, including John Charles Frémont's in 1848 and John W. Gunnison's in 1853). They had made no contact with the Spanish, and had met groups of Indians of several tribes without encountering serious hostilities.

Soon after his arrival in Fort Smith, Long dispatched a report to Secretary of War Calhoun. He may not have accomplished all the idealistic goals that had been set out for him, but he had reason to be satisfied with the accomplishments of the ill-equipped band that had struggled through rough and largely unexplored country and returned safely.

$\mathcal{E}leven$

EPILOGUE

THE EXPEDITION REMAINED AT FORT SMITH for only a few days before proceeding to its final destination, Cape Girardeau, on the Mississippi, where the men had planned to meet Lieutenant James Graham with the steamboat *Western Engineer*. They were anxious to return east with their notes and specimens, and made few observations as they crossed Arkansas and Missouri by horseback on relatively well traveled trails and wagon roads. So the expedition may be said to have ended at Fort Smith. From there on the men traveled in several separate groups.

Bell, with Dougherty and Oakley, was the first to leave, on September 19. James, Swift, and Captain Stephen Kearny set out the next day to visit the hot springs of central Arkansas. (Kearny, who was visiting Fort Smith, was then a young man of twenty-six; he went on to become a general and governor of California. It was his dispute with John Charles Frémont that led to the latter's court-martial in 1846.) Long, Say, Peale, Seymour, Wilson, Adams, Duncan, and Sweney left on September 21 and joined Bell en route to Cape Girardeau.

Nights along the route were usually spent at the cabins of settlers. On the first night, some of the men were given feather beds, but, according to James, they "spent an unquiet and almost sleepless night, and arose on the following morning unrefreshed, and with a painful feeling of soreness in our bones, so great a change had the hunter's life produced upon our habits." Those who slept on the floor had a better night.

Much of the country they crossed was hilly and heavily wooded, primarily with oak, hickory, ash, and maple. At this season, the ground was covered with acorns, on which the settlers fattened their hogs. There were many contacts with Cherokee Indians, some of whom grew cotton and had black slaves. With the memories of the many sandy arroyos they had crossed only a few weeks earlier, the men were much impressed with the clear streams that flowed through the Ozarks.

James described the hot springs (now a national park) at some length. The springs had not been described in detail until 1804, when William Dunbar and George Hunter, sent by Jefferson to explore the Ouachita River, succeeded in reaching the springs. Stephen Long had visited the springs when he crossed Arkansas in 1818, and James's *Account* includes some of Long's observations. American holly (*Ilex opaca*) and cassine holly (*I. vomitoria*) grew near the springs. The leaves of the latter were often used as a substitute for tea, but caused vomiting if used to excess. In pine woods, a false foxglove (*Gerardia*) was "one of the most conspicuous objects." "The angelica tree . . . is common along the banks of the creek, and bending beneath its heavy clusters of purple fruit." These small, spiny trees (*Aralia spinosa*) are often called Hercules'-club or Devil's walking-stick. Several kinds of ferns also grew among the rocks.

James and his companions returned to the Arkansas at Little Rock, then a village of six or eight houses. In nearby swamps grew cypress trees, "imparting a gloomy and unpromising aspect to the country. . . . In the cypress swamps, few other trees and no bushes are to be seen, and the innumerable conic excrescences called knees, which spring up from the roots, resemble monuments in a church-yard, and give a gloomy and peculiar aspect to the scenery."

EPILOGUE

Titian Peale, sketch of a cypress tree, at Cape Girardeau, Missouri, on or about October 12, 1820. (From the sketchbooks of Titian Ramsay Peale, Yale University Art Gallery, gift of Ramsay MacMullen, M.A.H. 1967)

EJ: On the 12th October the exploring party were all assembled at Cape Girardeau. Lieutenant Graham, with the steamboat Western Engineer, had arrived a day or two before from St. Louis; having delayed there some time subsequent to his return from the Upper Mississippi. In the discharge of the duties on which he had been ordered, [he] and all his party had suffered severely from bilious and intermitting fever.

Within a few days, most of the group assembled at Cape Girardeau had succumbed to "intermitting fever"—alternating periods of chills and fever—evidently malaria. James attributed the attacks to their having breathed "the impure and offensive atmosphere of the Arkansa bottoms" at Fort Smith. This was, of course, many years before mosquitoes were incriminated as the vectors of malaria. A common treatment was "administering large draughts of whiskey and black pepper," but this was "productive of much mischief." "Peruvian bark" (quinine) was also used, but often "injudiciously."

Cape Girardeau at this time consisted of about twenty log cabins and one or two brick buildings—"a miserable and forsaken little town," James wrote in his diary. The streets were gullied, and some were overgrown with thickets of ironweed and thistles "as to resemble small forests." The country near the town was forested with oaks, tulip trees, and tupelos, with an undergrowth of hazel and Hercules'-club. South of town were cypress swamps "extending with little interruption far to the south."

Long, though still suffering from malaria, soon left overland for Washington, as did Bell. Bell continued his journal as he crossed Tennessee and Virginia by horseback through generally pleasant autumn weather.

NOVEMBER 20. JB: At length, I arrived on the sumit opposite and in view of George Town, and the *Capitol* in Washington city over which was flying the *Star spangled banner*, the house being in Session. It is impossible to describe my feelings. I halted and dismounted from my horse, to contemplate the scene. Six years ago, I had witnessed the Capitol in flames fired by the hand of my countries foe, Phoenix-like it had risen from the ashes with ten fold splendor. [Bell had served in the War of 1812; the Capitol had been burned in 1814.] Four months ago, I had turned my back to the snow caped mountains, my face toward home, since then, I had encountered many hardships, fatigue, privations and famine, had seen the savage of the wilderness in his most native wildness and with all the vicitudes attending our tour have enjoyed health, and thanks to the Great Spirit have returned safe.

Bell calculated that he had traveled 936 miles from Cape Girardeau to Washington; his total distance since he left West Point on May 20 was 5,325 miles. On November 25, he was interviewed by a reporter from a Washington newspaper, the *National Intelligencer*. The reporter was most impressed by the fact that the explorers had met with Indian tribes "who were ignorant, not only of the existence of the People of the United States, but of the existence of a race of White People!"

EPILOGUE

On November 1, Say, Peale, Seymour, and Graham descended the Mississippi to New Orleans by steamboat. Say was not quite through finding new animals. During the trip, he discovered a lizard no more than four inches long, with a tail another two and a half inches, brown in color but with a darker streak along the sides. In a footnote, he named it *Scincus lateralis*. This is the brown skink, now called *Leiolopisma laterale*. These short-legged skinks scuttle about the ground in leaf litter and have been little studied since Say's discovery nearly two centuries ago.

From New Orleans, the four men traveled by ship to Philadelphia, arriving in January, 1821. A few weeks later, their specimens arrived, and soon Say and Peale were busy unpacking them. Peale set to work mounting specimens and making sketches. On August 29, 1821, Say wrote to his friend John Melsheimer, describing some of his experiences on the expedition as well as his present activities.

> I experienced much difficulty in preserving the insects which I collected, many of them are interesting, though they are not numerous. The Secretary of War has ordered our collections to be deposited in the Philadelphia museum subject to his orders, an arrangement which was anticipated from the commencement of our expedition. I am now engaged in describing the new species of which I find there are many amongst them.

Edwin James did not return to Philadelphia until the autumn of 1821. After leaving Cape Girardeau, he traveled for a time with Lieutenant William Swift, who was now in charge of the *Western Engineer*, but Swift decided to leave the steamboat for the winter at the mouth of the Cumberland River and go east by horseback. James, short of funds and still indisposed, decided to spend the winter in Smithland, Kentucky. Although he used the time to write up some of his data, it was largely, according to Roger Nichols and Patrick Halley, a winter of "sickness and despondency." How thorough was his depression is revealed in a letter dated October 26, 1821, that he posted to his brother from Cape Girardeau.

> I am full of complaining and bitterness against Maj. Long on account of the manner in which he has conducted the Expedition. . . . We have traveled near 2000 miles through an unexplored and highly interesting country and have returned home almost as much a stranger to it as before. I have been allowed neither time to examine and collect or means to transport plants and minerals. We have been hurried through the country as if our sole object had been, as it was expressed in the orders which we received at starting "to bring the expedition to as speedy a termination as possible."

In the spring James borrowed money and went on, but his troubles were not over. In Maryland his trunk was stolen, and although most of his possessions were recovered, the money was gone and he had to borrow again to reach Philadelphia.

After his arrival in Washington, Long prepared a formal report of the expedition for Secretary Calhoun. In it he reviewed the itinerary and provided a description of the country and the Indian tribes encountered. According to Long, Bell was busy preparing his journal for submittal to Calhoun, but there is no evidence that he did so. As noted earlier, Bell's journal was lost for many years before being rediscovered and published in 1957.

Long persuaded Calhoun to continue to pay the naturalists per diem while they worked on their notes and collections. He also obtained permission for them to prepare an account of their experiences for the general public. Long, Say, and James were given a small office at government expense, but had to supply their own paper and other supplies. On May 9, 1822, Say again wrote to John Melsheimer.

> I [have been] busily occupied with a portion of the labour of compiling the narrative of our journey to the Rocky Mountains. In addition to contributing my aid in the ordinary diatribe of the work, it falls to my lot to describe the new Quadripeds, birds & reptiles which we met with, as well as to give an account, both moral and physical, of the natives of the country through which we passed. The arranging and recording of the Meterological observations, made chiefly by myself, also fall to my share of the duties, though the general narrative is written by our companion Dr James. . . .

The descriptions of the animals and plants that appeared in the *Account* were, of course, written in Philadelphia on the basis of notes made in the field and whatever specimens the naturalists were able to bring back. One wonders to what extent the descriptions might have been improved or expanded had not many of Say's notes and belongings been stolen on the return along the Arkansas River. However, a good many specimens did survive the long trip east, as James reported at the conclusion of his narrative.

> EJ: Most of the collections made on this expedition have arrived at Philadelphia, and are in good preservation; they comprise, among other things, more than sixty prepared skins of new or rare animals. Several thousand insects, seven or eight hundred of which are probably new. . . . The herbarium con-

tains between four and five hundred species of plants new to the Flora of the United States, and many of them supposed to be undescribed.

Why the final preparation of the *Account* fell to James is unclear. Perhaps it was because he was not otherwise occupied. Say had been appointed curator of the American Philosophical Society and professor of natural history at the University of Pennsylvania. Long and his family had been ill from time to time, and there were duties connected with his position in the army. Say described numerous vertebrate animals in the *Account*, saving the insects for various other publications (Appendix 2). James described only thirteen species of plants in his narrative. Most of the plants were turned over to John Torrey, who described many of them in a series of three articles published from 1824 through 1827. Torrey recognized that many of the plants discovered by James had in fact been found a year earlier by Thomas Nuttall on his trip up the lower Arkansas basin. Although Nuttall did not publish his descriptions until 1835, Torrey felt obliged to defer to Nuttall as the initial discoverer. Thus he described only about seventy-five of the several hundred new species brought back by James (Appendix 3).

James, with help from Say, Long, and doubtless Peale and Seymour, completed the manuscript in mid-1822. The report reveals many evidences of the haste in which it was prepared. It was published by Carey and Lea of Philadelphia in two volumes. A few months later, a London publisher put out a three-volume edition. The published report is dedicated to "the Honourable J. C. Calhoun, Secretary of War . . . as a grateful acknowledgment of his indulgence and patronage."

To the narrative of the expedition were appended Say's vocabularies of Indian languages, various calculations of longitude and latitude, a day-by-day account of weather conditions throughout the trip, and Long's report to Calhoun as well as his map. Several of Seymour's illustrations were also included. In a foreword, James described the compilers' objectives in preparing the report and attempted to justify the expedition's failure to accomplish all that was expected of it.

> EJ: In selecting from a large mass of notes and journals the materials of the following volumes, our design has been to present a compendious account of the labours of the Exploring Party, and of such of their discoveries as were thought likely to gratify a liberal curiosity. . . . [W]e hope to have contributed something towards a more thorough acquaintance with the Aborigines of our country. In other parts of our narrative . . . we have turned our attention towards the phenomena of nature, to the varied and beautiful productions of animal and vegetable life, and to the . . . features of the inorganic creation. If in this attempt we have failed to produce any thing to amuse or instruct, the deficiency is in ourselves. . . .

> It will be perceived that the travels and researches of the Expedition, have been far less extensive than those contemplated in the . . . orders:—the state of the national finances, during the year 1821, having called for retrenchments in all expenditures of a public nature,—the means necessary for the farther prosecution of the objects of the Expedition, were accordingly withheld.

For many readers, the narrative provided a rewarding first impression of the High Plains and the Front Range of the Rockies. Reviews were mixed, and Long was frequently criticized for not having found the sources of the South Platte and the Arkansas and for not having found the Red River at all. But the *North American Review* rose to his defense (probably from the pen of its editor, Edward Everett).

> Detestable parsimony! . . . [T]he country, which had grown and is growing in wealth and prosperity beyond any

other and all other nations, too poor to pay a few gentlemen and soldiers for exploring its mighty rivers, and taking
possession of the empires, which Providence has called it
to govern! . . . Poor, indeed, we are in spirit, if not in finance, if we will not afford to pay the expense of making
an inventory of the glorious inheritance we are called to
possess. England, staggering and sinking under her bur-
dens, can fit out her noble expeditions. . . . France has her
intrepid naturalists. . . . [but] we cannot find a small party
of discovery in powder and ball enough to hunt withal, or
blankets and strouding to trade with the Indians. . . .

William H. Goetzmann speaks of historical commentary centering
on the expedition members' supposed "careless and seemingly cav-
alier interest in actual discovery," often attributed to the government's niggardly support of their efforts. This may explain, to a
degree, Long's decisions not to follow the South Platte and the Arkansas to their sources. But it is difficult to find, in the final report
of the expedition, very much evidence of carelessness or of lack of
enthusiasm for discovery.

It seems remarkable that even though the Long Expedition encountered Indians of several tribes, there were no unpleasant inci-
dents other than minor ones. In 1811 and 1812, Jean Baptiste
Champlain and many of his party had been killed by Arapahos, and
two years after the Long Expedition returned east, the Arikaras am-
bushed William Ashley's party of trappers, killing fourteen and effectively closing the Missouri for a time. Doubtless Long was blessed
with good luck, but some credit is due to his and Bell's diplomacy
and to the know-how of their guides and interpreters: Bijeau, Ledoux, and Julien.

Long was criticized for speaking of parts of the West as "desert"
when a better evaluation was expected. But Long's opinion should
not be rejected hastily. It was, as Goetzmann remarks, "an honest

and significant assessment of the plains area as it appeared to his generation and those which followed down to the Civil War." As late as 1957, historian Walter Prescott Webb exclaimed that "the heart of the West *is* a desert, unqualified and absolute. . . . If we do not understand the West it is because we perversely refuse to recognize this fact; we do not want the desert to be there. We prefer to loiter on its edges, skirt it, avoid it, and even deny it." To Webb, the eight states from Montana and Idaho to New Mexico and Arizona were the "desert states," and all other land west of the hundredth meridian was "desert rim." "At the heart of the desert," he added, "there is no drought, there is only occasional mitigation of dryness."

According to historian Donald Worster, "For almost everyone who has come into this country in modern times the land of the West has jolted the mind and tried the body. Very little of it has seemed designed for human ease. Even in these days of fast automobiles, the high plains are a trial of patience and a defiance to occupation."

Americans much prefer to think of the West as a garden rather than a desert as they continue to drain the aquifers and dam the last few streams, ever devoted to green lawns and golf courses as well as to wasteful methods of irrigation. According to Richard Beidleman, a planning document produced by the state of Colorado quite recently had described parts of the state as semiarid; but upon revision, the word "semiarid" was removed.

In writing about the High Plains, Richard Dillon has warned,

Much of the land is marginal in terms of agriculture and some of it will not even tolerate heavy grazing. Edwin James and Stephen Long were not false prophets, entirely. But, unfortunately, it took the Dust Bowl of the 1930s to prove it. . . . The spectre of the Dust Bowl haunts the West even today for it proved, for all time, that there lurks behind the myth of the Great American Desert of the 1820s a frightening amount of reality.

That the West is today thriving, at least in spheres surrounding its several "oasis cities," is attributable to herculean efforts that Long and his men could not have imagined in their wildest dreams. According to figures cited by Worster, by 1976 the federal Bureau of Reclamation operated 320 water-storage reservoirs, 344 diversion dams, 14,400 miles of canals, 900 miles of pipelines, 205 miles of tunnels, 145 pumping plants, and 16,240 miles of transmission lines. To that must be added the accomplishments of the Army Corps of Engineers and of state and local governments and private companies. All of this at a fantastic cost measured not only in dollars, but in the lives of a host of plants and animals whose habitats have been vastly altered or destroyed altogether. The native flora and fauna had evolved over millions of years to fruitfully occupy this semiarid land, but they have in considerable part been replaced, in a few decades, by an engineered landscape that will last as long as the United States remains peaceful and prosperous, and as long as population pressure does not empty the last well or drain the last stream dry.

Stephen Long's name, when it is remembered at all today, is almost invariably linked with the phrase "Great Desert" and with his failure to fulfill all the unreasonable goals that had been set for him. Yet Long was the first to persuade the government to include trained scientists on ventures through the West. The shorter expeditions he made before 1819 were important in establishing the American presence in parts of the Mississippi basin, and his 1823 expedition into Minnesota and Lake Superior (which also included scientists) added much to knowledge of those areas. Long's efforts as an explorer filled in a major gap between the earlier trips of Lewis and Clark and of Pike and the era of the fur traders, which was already in decline when Frémont, "pathfinder of the West," began the first of his expeditions in 1842.

On the Long Expedition, the first exploration of the West that included trained scientists, it was inevitable that the description and naming of new species of plants and animals be given priority—and, in any case, at that time biological science had scarcely advanced beyond that stage. Say and James applied themselves to the tasks at hand to the best of their abilities, and they saw to it that their findings became known to the public. Say described, by my count, 13 new mammals, 13 birds, 12 reptiles and amphibians, 4 arachnids and crustaceans, and more than 150 insects. To be sure, some of the names he proposed have fallen into disuse, either because the specimens were lost and his descriptions were vague, or because he inadvertently described as new species some that had been described by someone else. But he nevertheless provided an important first step in coming to grips with the previously almost wholly unknown fauna of the High Plains and the Rockies.

Of the plants collected by James, about 140 species were described as new by James, Torrey, Gray, Bentham, and others. James was the first to collect specimens in the rich alpine flora of the Rockies during his ascent of Pike's Peak—surely the high point of the expedition in every sense. In the *Transactions of the American Philosophical Society* for 1825, James provided a listing of many of the plants he had collected on the expedition. Several hundred species are listed, but unfortunately he did not usually state the locality in which each was taken, and he omitted the species yet to be described, so his list is of limited value.

The descriptions of geologic features, landforms, and summer weather along the route all added to knowledge of this vast new addition to the territory of the United States, and the map that Long prepared became, in William H. Goetzmann's phrasing, "a landmark of American cartography." Long made several errors and false assumptions on his map and often plotted his campsites incorrectly;

nevertheless, it became the standard map of the West until Fré-mont's surveys in the 1840s.

Long has been criticized for errors in his mapping, and the nat-uralists for not stating precisely where their specimens were col-lected. But much of this imprecision was inevitable. There were few known landmarks other than the Rocky Mountains, Long's and Pike's Peaks, and several rivers and streams. Sometimes they misi-dentified landmarks, as in the case of the Red River. Often the men simply did not know where they were except in a most general sense. Say was often led to say that specimens were taken "near the Rocky Mountains," or James that a plant was found "30 day's journey from the mountains."

The illustrations by Seymour that were included in the *Account* provided Easterners with their first visual images of a world vastly different from the shady, well-watered forests they were accustomed to, as well as their first glimpse of members of several tribes of western Indians. The narrative could also be read as an adventure story, and even today it stands as an important chronicle of a time when the land was empty of all but Indians and wildlife. Many a present-day naturalist might wish that he could encounter some of the birds and mammals that the expedition's members took for granted—not only free-ranging bison and wolves, but Carolina parakeets, passenger pi-geons, and ivory-billed woodpeckers!

Nearly all the zoological specimens collected on the expedition have since been lost. The Philadelphia Museum was moved from building to building until 1850, when much of the collection was sold to P. T. Barnum to be displayed in his American Museum in New York. Barnum's museum was destroyed by fire in 1865. Say took most of his insect specimens to New Harmony, Indiana, when he moved there in 1825. After his death, they were returned to the Philadelphia Academy of Natural Sciences. In 1836, they were bor-rowed by Massachusetts entomologist Thaddeus Harris, who found them in "a deplorable condition, most of the pins having become loose, the labels detached, & the insects themselves without heads, antennae, & legs, or devoured by destructive larvae & ground to powder by the perilous shakings they have received." Some of Say's specimens found their way to Harvard's Museum of Comparative Zoology, where they can still be seen and studied. A few of the species that Say collected on the Long Expedition are represented in this material. Fortunately, the plants that had been sent to John Torrey have been better cared for, and most of them can still be found in the herbarium of the New York Botanical Garden.

It was more than a decade before naturalists once again penetrated as far as the Rockies in the search for plants and animals. Thanks to the efforts of George Catlin and Karl Bodmer, in the 1830s, landscapes and Native inhabitants of the northern plains and the foothills became known through a series of brilliant paintings and the collection of Native artifacts. Bodmer's patron, the German prince Alexander Philipp Maximilian, was willing to endure incredible hardships to study America's inhabitants, both human and animal. But it was not until the second expedition of Nathaniel Wyeth, in 1834/1835, that two well-qualified naturalists traveled deeply in the West—indeed, to the Pacific. The two were John Kirk Townsend and the indefatigable Thomas Nuttall. In 1844, Audubon was collecting mammals on the upper Missouri and bemoaning the fate of the bison—as Edwin James had done twenty-four years earlier.

That the Long Expedition was largely considered a success is demonstrated by the fact that Long was asked, only three years later, to undertake a similar exploration of St. Peter's River and the Red River of the North, in Minnesota and Manitoba, returning via the Great Lakes. He was again accompanied by Say and Seymour. James was invited to participate, but he did not learn of the invitation until the expedition had left Philadelphia in April 1823. Say was therefore asked to collect plants as well as animals, even though he was not an enthusiastic botanist. Some of his collections were lost in transit, but he did bring back many insects as well as much information on the Indians the explorers encountered.

This was Stephen Long's last expedition of exploration. As a major and later a colonel in the Corps of Topographical Engineers, he was assigned to improving the navigability of rivers and to planning railroads. In 1829, he published a manual of railroad construction and, in 1836, a booklet on bridge building that included several bridges of his own design. Later be became involved in hospital and steamboat construction. At the time of the Civil War, he was elevated to chief of the Corps of Topographical Engineers. In 1863, he retired, at the age of seventy-nine and after having spent nearly fifty years in the army. He died a year later.

Long's military assistants on the 1819/1820 expedition, James Graham and William Swift, both went on to distinguished careers. Graham was involved in surveying the boundary with Mexico; Mount Graham in Arizona is named for him. Later he discovered lunar tides on the Great Lakes. By 1863, he had reached the rank of colonel. He died in 1865 while inspecting a seawall in Massachusetts during a storm. Swift resigned from the army in 1849 and became president of the Philadelphia, Wilmington, and Baltimore Railroad. He was an art lover and friend of James MacNeill Whistler. As a businessman, he was said to be shrewd but always scupulous and honest. He died in 1885.

John Bell lived less than five years after returning to Washington. He was first posted to Florida, where a dispute with a junior officer brought about his court-martial in 1821. He was found guilty of conduct unbecoming an officer, but his sentence was remitted a few months later. After serving for two years in Georgia and South Carolina, he resigned from the army because of poor health. He was only about forty years of age when he died in 1825.

Thomas Say had scarcely had time to help with the completion of the *Account* of the 1819/1820 expedition before he was off on his second trip with Long. Only two years after his return from this trip, he left Philadelphia permanently to participate in Robert Owen's utopian experiment at New Harmony, Indiana. In 1827 and 1828, he made his last major field trip, this one to Mexico with his patron, William Maclure. Before leaving Philadelphia, he had completed part of his *American Entomology* and prepared several other papers for publication. He also assisted Charles Lucien Bonaparte with his supplement to Alexander Wilson's *American Ornithology*. At New Harmony, Say continued to publish extensively on insects and shells, though he became more and more involved with administrative duties. In 1827, he eloped with his artist, Lucy Sistaire, but he lived for only seven more years, dying at the age of forty-seven with his life's work far from complete.

Say is often called "the father of American entomology," an appellation that is both too broad and too restrictive: too broad in that he was strictly a systematist, with little interest in insect life histories or control; too restrictive in that his contributions to conchology, to vertebrate zoology, and to Indian ethnology were substantial. Say corresponded with zoologists in this country and abroad, and he was elected a foreign member of the prestigious Linnean Society of London as well as the Zoological Society of London. After his death, his friend George Ord spoke of him at a meeting of the American Philosophical Society:

> Although on the score of Mr. Say's literary acquirements there may be some diversity of opinion, yet there can be but one sentiment with regard to his industry, his zeal, and the extent of his knowledge of natural history, particularly of that class of zoology to which he was most attached, Entomology. His discoveries of new species of insects were, perhaps, greater than ever had been made by a single individual.

Titian Peale long outlived Say and all the other naturalists he had known as a youth, dying in 1885 at the age of eighty-six. After his return from the West, he provided illustrations for Say's *American Entomology* and for Bonaparte's research on American

Titian Peale, pencil sketch of an Indian on horseback. (American Philosophical Society)

birds. In 1831, he joined an expedition to Colombia, and from 1838 to 1842 he was a member of the Wilkes Expedition, traveling throughout the southern oceans and along the west coast of North America on the *Peacock*. From 1849 to 1872, he was an examiner in the Patent Office in Washington, where he helped found the Washington Philosophical Society. His spare time was spent in photography and in preparing oil paintings based on sketches he had made on his various expeditions. He had lived an adventurous life and left behind images that had long since become part of the country's vision of the West, particularly his sketches of Indians hunting bison, which became well known through the engravings of Currier and Ives and others.

Samuel Seymour remains an enigmatic figure, only occasionally mentioned in James's *Account* and in William H. Keating's *Narrative of an Expedition to the Source of St. Peter's River*. He is said to have completed some 150 illustrations from the 1819/1820 expedition, but only about 25 remain extant, and only 8 were included in the

Account. Keating's *Narrative* included eleven of Seymour's renditions of Indians and scenery. Following the latter expedition, Long paid Seymour $2 a day for three months to complete his work, but after that Seymour seems to have vanished without a trace. For a time, some of his paintings hung on the walls of the Philadelphia Museum, but they disappeared when the museum's holdings were sold at auction in 1850. Of Seymour's work, the Goetzmanns have this to say:

> His work seems spare, contrived and even clumsy to us today. But he was an artist in the topographical tradition that began to be in vogue in the mid-eighteenth century when exploring expeditions circled the globe. . . . The duty of the topographical artist was to render landforms as exactly as possible . . . but as Seymour's renditions indicate, the artist's feelings of awe and wonder at the moment of viewing inevitably allowed his emotions to give form and character to the pictures.

It should be pointed out that Seymour's paintings are so faithful to the landscape that many of the landforms he painted can be recognized today. Indeed, people attempting to follow the route of the expedition may—and often do—use the paintings as a guide. But it should be pointed out that engravers sometimes took liberties with Seymour's paintings. In his *View of the Chasm Through which the Platte Issues from the Rocky Mountains,* Seymour imaginatively included an Indian and a member of the expedition. Perhaps hoping to heighten the sense of untamed wilderness, the engravers for the *Account* omitted these figures. Other engravings also were less than perfect renditions of the originals. Copies of illustrations from the *Account* often appeared in other publications, sometimes with still further modifications. The realism practiced by the exploring party's artists was often deemed too bland for the growing romanticism of the nineteenth century. Catlin's and Bodmer's paintings of Indians, made in the 1830s, and Albert Bierstadt's landscapes, made in the 1860s, soon overshadowed the modest contributions of Peale and Seymour. But they were the first!

Although Edwin James was in many ways the "star" of the Long Expedition, he gradually drifted away from botany and finally from science altogether. Shortly after his return east, he wrote to Calhoun asking that he be assigned to the Medical Department of the army. He served for three years at Fort Crawford, near Prairie du Chien, Wisconsin. He tried to launch an expedition to the west coast, but Calhoun felt that he lacked sufficient experience, and nothing came of his plans. On a leave of absence in the East, in 1827, he married; after that he was reassigned to posts in northern Michigan. At these posts, he had much contact with the Ojibwa Indians, and he translated the New Testament into their language. One of his converts was John Tanner, who had been stolen from his home as a child by the Indians and reared among them. Tanner married a squaw and succumbed to whiskey, but James induced him to reform. In 1830 he published a biography of Tanner. He also prepared grammars and spelling books in Indian languages.

In 1834 James resigned from the army and moved to Albany, New York, where for a time he edited the *Temperance Herald and Journal*. A few years later, he moved to a farm near Burlington, Iowa. For a time he served as an Indian agent in Council Bluffs, but he became bitter about the lack of sympathy and support for Native Americans and soon resigned and returned to his farm. He was an ardent abolitionist, and his home served as a way station on the Underground Railroad for runaway slaves. He became something of an eccentric and a mystic, and in an 1854 letter to John Torrey he confessed that he had become immersed in "the chill and foggy domains of theology." Yet he dreamed of one day going forth again "to gather weeds and stones and rubbish . . . and perhaps drop this life-wearied body beside some solitary stream in the wilderness." In a letter to botanist Charles Parry, in 1859, James wrote, "I became a settler in Iowa twenty-two years ago and of course have seen great changes. The locomotive engine and the railroad car scour the plain in place of the wolf and the curlew. Mayweed and dog fennel, stink weed and mullein have taken the place of purple flox and the mocassin flower."

Late in life, James received a letter from Stephen Long, suggesting that he write a book reviewing Long's expeditions. He began work on it, but apparently never got very far. On October 25, 1861, he fell from a load of wood on a wagon and was run over by the wheels of the wagon. He died a few days later.

For all these men, the memory of the days when they struggled, half-starved, along the foothills of the Rockies, recording plants, animals, and landforms never before seen by western people, must have seemed a bright flash of light in lives clouded with responsibilities less taxing but lacking the rewards of discovery and the purity of a new, untrammeled world.

$\mathcal{A}ppendix$ 1

ANIMALS (OTHER THAN INSECTS) DESCRIBED BY THOMAS SAY IN THE *ACCOUNT* OF THE LONG EXPEDITION, 1819—1820

MAMMALS

Canis latrans, **coyote**

Canis nubilus (now *C. lupus nubilus*), **prairie gray wolf**

Canis velox (now *Vulpes velox*), **swift fox**

Cervus macrotis (antedated by *C. hemionus* Rafinesque, now *Odocoileus hemionus*), **mule deer**

Sciurus grammurus (now *Spermophilus variegatus grammurus*), **rock squirrel**

Sciurus lateralis (now *Spermophilus lateralis*), **golden-mantled ground squirrel**

Sciurus macrurus (antedated by *S. niger rufiventre* Geoffroy St. Hilaire), **fox squirrel**

Sciurus quadrivittatus (now *Eutamias quadrivittatus*), **Colorado chipmunk**

Sorex brevicauda (now *Blarina brevicauda*), **short-tailed shrew**

ANIMALS DESCRIBED IN THE *ACCOUNT*

Sorex parvus (now *Cryptotis parva*), **least shrew**

Vespertilio arquatus (antedated by V. *fuscus* Beauvois, now *Eptesicus fuscus*), **big brown bat**

Vespertilio pruinosus (antedated by V. *cinereus* Beauvois, now *Lasiurus cinereus*), **hoary bat**

Vespertilio subulatus (now *Myotis subulatus*), **small-footed myotis**

BIRDS

Columba fasciata, **band-tailed pigeon**

Emberiza amoena (now *Passerina amoena*), **lazuli bunting**

Fringilla frontalis (now *Carpodacus mexicanus frontalis*), **house finch**

Fringilla grammaca (now *Chondestes grammacus*), **lark sparrow**

Fringilla psaltria (now *Carduelis psaltria*), **lesser goldfinch**

Hirundo lunifrons (antedated by H. *pyrrhonota* Vieillot), **cliff swallow**

Limosa scolopacea (now *Limnodromus scolopaceus*), **long-billed dowitcher**

Pelidna pectoralis (antedated by P. *melanotus* Vieillot, now *Calidris melanotos*), **pectoral sandpiper**

Sylvia bifasciata (antedated by S. *cerulea*, Wilson, now *Dendroica cerulea*), **cerulean warbler**

Sylvia celatus (now *Vermivora celata*), **orange-crowned warbler**

Tetrao obscurus (now *Dendragapus obscurus*), **blue grouse**

Troglodytes obsoleta (now *Salpinctes obsoletus*), **rock wren**

Tyrannus verticalis, **western kingbird**

REPTILES

Agama collaris (now *Crotaphytus collaris*), **eastern collared lizard**

Ameiva tesselata (now *Cnemidophorus tesselatus*), **checkered race runner**

Coluber flaviventris (now C. *constrictor flaviventris*), **blue racer**

Coluber obsoletus (now *Elaphe obsoleta*), **pilot black snake**

Coluber parietalis (now *Thamnophis sirtalis parietalis*), **red-sided garter snake**

Coluber proximus (now *Thamnophis sauritus proximus*), **western ribbon snake**

Coluber testaceus (now *Masticophus flagellum testaceus*), **western coachwhip**

Crotalus confluentus (antedated by C. *viridis* Rafinesque), **prairie rattlesnake**

Crotalus tergeminus (now *Sistrurus catenatus tergeminus*), **western massasauga**

Scincus lateralis (now *Leiolopisma laterale*), **brown skink**

AMPHIBIANS

Bufo cognatus, **great plains toad**

Triton lateralis (antedated by *Siren maculosus* Rafinesque, now *Necturus maculosus*), **mud puppy**

ARACHNIDS

Galeodes pallipes (now *Eremobates pallipes*), **a sunspider**

Galeodes subulata, opposite sex of G. *pallipes*, which preceded it on page and therefore has priority

Ixodes molestus (name never used because description was inadequate and specimen was lost), **a tick.**

CRUSTACEANS

Apus obtusus (now *Triops longicaudatus*), **a tadpole shrimp**

Appendix 2
INSECTS DESCRIBED BY THOMAS SAY
FROM SPECIMENS COLLECTED ON THE
LONG EXPEDITION, 1819–1820

Thomas Say chose not to include descriptions of insects in the *Account*, doubtless because it would have greatly increased its length and delayed publication. Rather, they were published from 1824 to 1835 in his book *American Entomology* and in the *Western Quarterly Reporter, Journal of the Academy of Natural Sciences of Philadelphia, Annals of the Lyceum of New York, Transactions of the American Philosophical Society,* and *Boston Journal of Natural History.* About 160 species were clearly described from the expedition, as indicated by Say's accompanying notes. Say also described many other species simply from "Arkansas," "Missouri," or "United States." Some of these may have been taken on the expedition, but they are not included here for lack of clear evidence. Say's publications were compiled, edited, and reprinted by John L. LeConte in 1859.

HYMENOPTERA (WASPS AND BEES)

Anomalon flavicornis (now *Thyreodon atricolor flavicorne*), **an ichneumon wasp**

INSECTS DESCRIBED IN THE *ACCOUNT*

Astata bicolor, **a digger wasp**

Epeolus quadrifasciatus (now *Triepeolus quadrifasciatus*), **a parasitic bee**

Evania unicolor (antedated by *E. appendigaster* Linnaeus), **an ensign wasp**

Mutilla contracta (name unused because of inadequate description and loss of specimen), **a velvet ant**

Mutilla quadriguttata (now *Dasymutilla quadriguttata*), **a velvet ant**

Philanthus canaliculatus (now *Eucerceris canaliculata*), **a digger wasp**

Philanthus zonatus (now *Eucerceris zonata*), **a digger wasp**

Pompilus formosus (now *Pepsis formosa*), **a spider wasp**

Pompilus terminatus (now *Cryptocheilus terminatus*), **a spider wasp**

Stizus grandis (now *Sphecius grandis*), **western cicada killer wasp**

Stizus renicinctus (now *Stizoides renicinctus*), **a parasitic digger wasp**

Tremex obsoletris (antedated by *T. columba* Linnaeus), **a wood wasp**

Tremex sericeus (antedated by *T. columba* Linnaeus), **a wood wasp**

DIPTERA (FLIES)

Anthrax irroratus, **a bee fly**

Chrysops quadrivittatus (now *Silvius quadrivittatus*), **a deer fly**

Dilophus stigmaterus, **a march fly**

Gonia frontosa, **a tachina fly**

Heleomyza 5-punctata (now *Suillia 5-punctata*), **a fungus fly**

Laphria fulvicauda (now *Andrenosoma fulvicauda*), **a robber fly**

Tabanus molestus, **a horse fly**

Zodion abdominalis (antedated by *Z. fulvifrons* Say), **a conopid fly**

ORTHOPTERA (CRICKETS AND GRASSHOPPERS)

Acheta exigua (now *Anaxipha exigua*), **Say's bush cricket**

Gryllus bivittatus (now *Melanoplus bivittatus*), **two-striped grasshopper**

Gryllus formosus (now *Tropidolophus formosus*), **great crested grasshopper**

Gryllus hirtipes (now *Acrolophitus hirtipes*), **crested-keel grasshopper**

Gryllus nubilus (now *Boopedon nubilum*), **boopee grasshopper**

Gryllus trifasciatus (now *Hadrotettix trifasciatus*), **three-banded range grasshopper**

HEMIPTERA (TRUE BUGS)

Acanthia interstitialis (antedated by A. *pallipes* Fabricius, now *Saldula pallipes*), **a shore bug**

Cercopis obtusa (now *Clastoptera obtusa*), **alder spittlebug**

Cicada aurifera (now *Tibicen aurifera*), **a cicada**

Cicada dorsata (now *Tibicen dorsata*), **a cicada**

Cicada parvula (antedated by C. *calliope* Walker, now *Melampsalta calliope*), **a cicada**

Cicada pruinosa (now *Tibicen pruinosa*), **a cicada**

Cicada synodica (now *Okanagana synodica*), **a cicada**

Coreus lateralis (now *Corizus lateralis*), **a coreid bug**

Cydnus spinifrons (now *Amnestus spinifrons*), **a burrower bug**

Delphax tricarinata (now *Stobaera tricarinata*), **a planthopper**

Flata bivittata (now *Acanalonia bivittata*), **a planthopper**

Flata stigmata (now *Cixius stigmata*), **a planthopper**

Fulgora sulcipes (now *Scolops sulcipes*), **a planthopper**

Lygaeus trivittatus (now *Leptocoris trivittatus*), **boxelder bug**

Reduvius spissipes (now *Apiomerus spissipes*), **an assassin bug**

Tettigonia limbata (antedated by *T. septentrionalis* Walker, now *Cuerna septentrionalis*), **a leafhopper**

Tettigonia obliqua (now *Erythroneura obliqua*), **a leafhopper**

Tettigonia 8-lineata (now *Gyponana 8-lineata*), **a leafhopper**

COLEOPTERA (BEETLES)

Say described more than 100 species of beetles from the Long Expedition. Many of his species have now been removed to more modern genera, but rather than indicate this I have chosen to retain his original names and to group the species according to the family to which each belongs. The study of beetles is a world in itself; all told, there are hundreds of families and untold tens of thousands of species. Beetles were favorites of Say, and they are easier to collect and preserve than some other insects.

Bruchidae (seed beetles): **Bruchus discoideus**

Buprestidae (metallic wood borers): **Buprestis atropurpureus,B. campestris, B. confluenta, B. pusilla**

Cantharidae (soldier beetles): **Cantharis jactata, C. ligata**

Carabidae (ground beetles): **Bembidium inaequalis, B. sigillare, Brachinus cyanipennis, B. stygicornis, Calosoma obsoleta, Cymindis laticollis, Feronia constricta**

Cerambycidae (longhorned beetles): **Callidium amoenum, C. fulvipenne, C. 6-fasciatum, Cerambyx solitarius, Leptura bivittata, Moneilema annulata, Prionus emarginata, P. palparis, Saperda pergrata, Stenocorus mucronatus**

Chrysomelidae (leaf beetles): **Altica 5-vittata, Chrysomela auripennis, C. basilaris, Colaspis dubiosa, C. interrupta,Cryptocephalus bivittatus, C. confluens, Donacia aequalis, Doryphora 10-lineata,**

Galleruca dorsata, G. longicornis, G. puncticollis, G. tricincta, Hispa lateralis, Imatidium 17-punctatum, Lema trivittata

Cicindelidae (tiger beetles): *Amblycheila cylindriformis Cicindela limbata, C. obsoleta, C. pulchra*

Cleridae (checkered beetles): *Trichodes ornatus*

Coccinellidae (lady beetles): *Coccinella humeralis*

Colydiidae (cylindrical bark beetles): *Synchita 4-guttata*

Curculionidae (weevils): *Brachycerus humeralis, Calandra compressirostra, Liparus imbricatus, L. vittatus, Rhynchaenus lineaticollis*

Dytiscidae (predaceous diving beetles): *Colymbetes venustus, Hydroporus undulatus*

Elateridae (click beetles): *Elater rectangularis*

Heteroceridae (variegated mud-loving beetles): *Heterocerus pallidus, H. pusillus*

Histeridae (hister beetles): *Hister bifidus, H. sedecimstriatus*

Hydrophilidae (water scavenger beetles): *Helophorus lineatus, Hydrophilus triangularis, Sphaeridium apicalis*

Lathridiidae (minute brown scavenger beetles): *Lathridium 8-dentatus*

Lucanidae (stag beetles): *Lucanus placidus, Platycerus securidens*

Lycidae (net-winged beetles): *Lycus sanguinipennis, L. terminalis*

Meloidae (blister beetles): *Lytta albida, L. nuttalli, Meloe conferta, Nemognatha atripennis, N. minima*

Melyridae (soft-winged flower beetles): *Malachius bipunctatus*

Oedemeridae (false blister beetles): *Oedemera ruficollis*

Phalacridae (shining flower beetles): *Phalacrus penicillatus*

Scarabaeidae (scarab beetles): *Aphodius concavus, A. femoralis, A. strigatus, Ateuchus nigricornis, A. obsoletus, Cetonia barbata, Geotrupes excrementi, G. filicornis, Melolontha lanceolata, M. 10-lineata, M. pilosicollis, M. sericea, Trox scutellaris*

INSECTS DESCRIBED IN THE *ACCOUNT*

Silphidae (carrion beetles): **Agathidium pallidum, Catops basilaris**

Staphylinidae (rove beetles): **Anthophagus brunneus, Omalium marginatum, Oxytelus fasciatus, O. melanocephalus, O. pallipennis, Stenus quadripunctatus, Tachinus atricaudatus**

Tenebrionidae (darkling beetles): **Akis muricata, Asida anastomosis, A. opaca, A. polita, Blaps acuta, B. carbonaria, B.extricata, B. obscura, B. obsoleta, B. suturalis, Diaperis bifasciata, Opatrum pullum, Pimelia rotunda, Zophosis reticulata**

Appendix 3
PLANTS NEWLY DISCOVERED AND DESCRIBED FROM THE LONG EXPEDITION,
1820

Edwin James brought back specimens of about 700 species of plants during his time with the Long Expedition as it traveled from the Missouri to Fort Smith, Arkansas, from June to early September 1820. Of these, about 140 proved to be new to science and were subsequently described and named by James himself (13), by John Torrey (76), by Torrey and Asa Gray together (13), and by the English botanist George Bentham (5). These are listed here. The remaining few were described by several other individuals, about a dozen of them by Constantine Rafinesque. Rafinesque's names have largely passed into obscurity and are not listed. A notable exception is *Cercocarpus montanus* (mountain-mahogany), based on specimens that James collected probably in present Jefferson County, Colorado. This is an abundant shrub in the foothills of the Front Range and provides important browse for deer and elk.

George Goodman and Cheryl Lawson's, *Retracing Major Stephen H. Long's 1820 Expedition* should be consulted for a fuller discussion of the plants. This listing relies heavily on their book.

PLANTS DESCRIBED IN THE *ACCOUNT*

PLANTS DESCRIBED BY EDWIN JAMES (1823, 1825)

Aquilegia coerulea, **Colorado blue columbine**

Argemone alba (now *A. polyanthemos*), **prickly poppy**

Cucumeris perennis (antedated by *Cucurbita foetidissima*), **buffalo-gourd**

Gaura mollis, **velvet-leaf gaura**

Geranium intermedium, **a wild geranium**

Hieracium runcinatum (now *Crepis runcinata*), **hawksbeard**

Pinus flexilis, **limber pine**

Populus angustifolia, **narrowleaf cottonwood**

Prenanthes runcinata (now *Stephanomeria pauciflora*), **wirelettuce**

Ranunculus amphibius, **a water crowfoot**

Rudbeckia tagetes (now *Ratibida tagetes*), **prairie coneflower**

Solanum hirsutum (now *S. triflorum*), **cut-leaved nightshade**

Stanleya integrifolia (now *S. pinnata integrifolia*), **prince's plume**

PLANTS DESCRIBED BY JOHN TORREY (1824–1827)

Acer glabrum, **Rocky Mountain maple**

Agrostis airoides (now *Sporobolus airoides*), **alkali sacaton**

Agrostis caespitosa (now *Muhlenbergia torreyi*), **a muhly grass**

Agrostis cryptandra (now *Sporobolus cryptandrus*), **sand dropseed**

Andropogon glaucum (now *A. saccharoides*), **silver bluestem grass**

Androsace carinata, **rock jasmine**

Arenaria obtusa (now *A. obtusiloba*), **alpine sandwort**

Aristida fasciculata (antedated by *A. adscensionis*), **three-awn grass**

Artemisia filifolia, **silvery wormwood or sand sagebrush**

Asclepias jamesii (antedated by A. *latifolia*), **broadleaved milkweed**

Asclepias obtusifolia latifolia (now A. *latifolia*), **broadleaved milkweed**

Asclepias speciosa, **showy milkweed**

Astragalus mollissimus, **woolly locoweed**

Bidens gracilis (now *Thelesperma megapotamica*), **greenthread**

Cantua longiflora (now *Ipomopsis longiflora*), **pink gilia**

Cantua pungens (now *Leptodactylon pungens*), **a gilia**

Capraria pusilla (now *Mimulus floribundus*), **many-flowered mimulus**

Carex jamesii (now C. *nebraskensis*), **a sedge**

Castilleja occidentalis, **western yellow paintbrush**

Celtis reticulata, **hackberry**

Chenopodium simplex, **a goosefoot**

Cleomella angustifolia, **cleomella**

Dalea formosa, **feather plume**

Darlingtonia intermedia (now *Desmanthus illinoensis*), **bundleflower**

Eriogonum tenellum, **a wild buckwheat**

Eriogonum umbellatum, **sulphur-flower**

Eryngium diffusum, **eryngo**

Frankenia jamesii, **James's frankenia**

Gaillardia pinnatifida, **a blanketflower**

Gaura parvifolia (now G. *coccinea parvifolia*), **a butterfly weed**

Gaura villosa, **hairy gaura**

Inula ericoides (now *Leucelene ericoides*), **white aster**

Ionidium lineare (now *Hybanthus linearis*), **green violet**

Krameria lanceolata, **krameria**

PLANTS DESCRIBED IN THE *ACCOUNT*

Lithospermum decumbens (now *L. caroliniensis*), **a puccoon**

Lupinus decumbens (antedated by *L. argenteus*), **silvery lupine**

Myosotis suffruticosa (now *Cryptantha cinerea*), **James's cryptantha**

Ornithogalum bracteatum (now *Lloydia serotina*), **alp lily**

Oxybaphus multiflorus (now *Mirabilis multiflora*), **Colorado four-o'clock**

Pectis angustifolia, **lemon-scented pectis**

Pentstemon alpina (now *Penstemon glaber alpinus*), **alpine penstemon**

Pentstemon ambiguum (now *Penstemon ambiguus*), **plains penstemon**

Petalostemum macrostachyum (now *Dalea cylindriceps*), **a prairie clover**

Phacelia integrifolia, **a scorpion weed**

Physalis lobata, **purple-flowered ground-cherry**

Plantago eriopoda, **redwool plantain**

Pleuraphis jamesii (now *Hilaria jamesii*), **galleta-grass**

Potentilla leucophylla (now *P. hippiana*), **woolly cinquefoil**

Potentilla nivalis (now *Geum rossii turbinatum*), **alpine avens**

Prenanthes pauciflora (now *Stephanomeria pauciflora*), **wirelettuce**

Prenanthes tenuifolia (now *Stephanomeria minor*), **small wirelettuce**

Primula angustifolia, **alpine primrose**

Prosopis glandulosa, **honey mesquite**

Psoralea jamesii (now *Dalea jamesii*), **James's prairie clover**

Pulmonaria alpina (now *Mertensia alpina*), **alpine bluebells**

Pulmonaria ciliata (now *Mertensia ciliata*), **tall bluebells**

Quercus undulata, **wavyleaf oak**

Rubus deliciosus, **Boulder raspberry**

Rubus idaeus (antedated by *R. occidentalis*), **black raspberry**

Saxifraga jamesii (now *Telesonix jamesii*), **James's saxifrage**

Scutellaria resinosa, **resinous skullcap**

Sedum lanceolatum, **stonecrop**

Sida stellata (now *Sphaeralcea angustifolia cuspidata*), **a globe mallow**

Solanum flavidum (now *S. elaeagnifolium*), **silverleaf nightshade**

Solanum jamesii, **wild potato**

Spiraea monogyna (now *Physocarpus monogynus*), **ninebark**

Stellaria jamesiana, **tuber starwort**

Stevia sphacelata (now *Palafoxia sphacelata*), **rayed palafoxia**

Stillingia sylvatica salicifolia, **queen's delight**

Teucrium laciniatum, **germander**

Tiarella bracteata (now *Heuchera bracteata*), **bracted alum-root**

Tragia ramosa, **branching tragia**

Trifolium nanum, **dwarf alpine clover**

Uniola stricta (now *Distichlis stricta*), **a saltgrass**

Vernonia altissima marginata (now *V. marginata*), **plains ironweed**

Zapania cuneifolia (now *Phyla cuneifolia*), **fogfruit**

PLANTS DESCRIBED BY JOHN TORREY AND ASA GRAY
(1838–1843)

Berlandiera incisa (antedated by *B. lyrata*), **lyreleaf greeneyes**

Crepis runcinata, **hawksbeard**

Cristatella jamesii (now *Polanisia jamesii*), **James's crestpetal**

Hoffmanseggia jamesii (now *Caesalpinia jamesii*), **James's rushpea**

Jamesia americana, **waxflower**

Linosyris pluriflora (now *Haplopappus pluriflorus*), **a goldenweed**

PLANTS DESCRIBED IN THE *ACCOUNT*

Malva involucrata (now *Callirhoe involucrata*), **purple poppy mallow**

Oenothera coronopifolia, **cut-leaved evening primrose**

Oenothera jamesii, **James's evening primrose**

Oenothera lavandulifolia (now *Calylophus lavandulifolius*), **lavenderleaf evening primrose**

Paronychia jamesii, **James's nailwort**

Stenotus pygmaeus (now *Haplopappus pygmaeus*), **a goldenweed**

Trifolium dasyphyllum, **whiproot clover**

PLANTS DESCRIBED BY GEORGE BENTHAM (1846–1856)

Chionophila jamesii, **snow-lover**

Eriogonum jamesii, **James's wild buckwheat**

Penstemon jamesii, **James's beardtongue**

Penstemon torreyi (now *P. barbatus torreyi*), **Torrey's beardtongue**

Synthyris plantaginea (now *Besseya plantaginea*), **kitten-tail**

BIBLIOGRAPHY

Beidleman, Richard G. "Edwin James: Pioneer Naturalist." *Horticulture* 44 (1966): 32–34.

———. "The 1820 Long Expedition." *American Zoologist* 26 (1986): 307–313.

Bell, John R. *The Journal of Captain John R. Bell, Official Journalist of the Stephen H. Long Expedition to the Rocky Mountains, 1820.* Edited by Harlin M. Fuller and Le Roy R. Hafen. The Far West and the Rockies Historical Series, Vol. 6. Glendale, Calif.: Clark, 1957.

Benson, Maxine. "Edwin James, Scientist, Linguist, Humanitarian." Ph.D. diss., University of Colorado, 1968.

——— ed. *From Pittsburgh to the Rocky Mountains: Major Stephen Long's Expedition, 1819–1820.* Golden, Colo.: Fulcrum, 1988.

Bentham, George. [Several taxonomic treatises.] In *Prodromus Systematis Naturalis Regni Vegetabilis,* edited by A. DeCandolle. Paris: Masson, 1845–1856.

Brackenridge, Henry Marie. *Views of Louisiana: The Journal of a Voyage up the Missouri River, In 1811* (1814). In *Early Western Travels, 1748–1846,* edited by Reuben Gold Thwaites, Vol. 6. Cleveland: Clark, 1904.

Bradbury, John. *Travels in the Interior of America in the Years 1809, 1810, and 1811,* (1819). In *Early Western Travels, 1748–1846,* edited by Reuben Gold Thwaites, vol. 5. Cleveland: Clark, 1904.

Chittenden, Hiram M. *The American Fur Trade of the Far West.* 2 vols. 1902. Reprint. Lincoln: University of Nebraska Press, 1986.

BIBLIOGRAPHY

Cutright, Paul Russell. *Lewis and Clark: Pioneering Naturalists.* 1969. Reprint. Lincoln: University of Nebraska Press, 1989.

Dillon, Richard. "Stephen Long's Great American Desert." *Proceedings of the American Philosophical Society* 111 (1967): 93–108.

Ewan, Joseph. *Rocky Mountain Naturalists.* Denver: University of Denver Press, 1950.

Ewers, John C. *Artists of the Old West.* Garden City, N.Y.: Doubleday, 1965.

Goetzmann, William H. *Army Exploration in the American West, 1803–1863.* New Haven: Yale University Press, 1959.

———. *Exploration and Empire: The Explorers and the Scientists in the Winning of the American West.* 1966. Reprint. New York: Norton, 1978.

Goetzmann, William H., and William N. Goetzmann. *The West of the Imagination.* New York: Norton. 1986.

Goodman, George J., and Cheryl A. Lawson. *Retracing Major Stephen H. Long's 1820 Expedition: The Itinerary and Botany.* Norman: University of Oklahoma Press, 1995.

Haltman, Kenneth. "Between Science and Art: Titian Ramsay Peale's Long Expedition Sketches, Newly Recovered at the State Historical Society of Iowa." *Palimpsest* 74 (1993): 62–81.

———. "Private Impressions and Public Views: Titian Ramsay Peale's Sketchbooks from the Long Expedition, 1819–1820." *Yale University Art Gallery Bulletin,* Spring 1989, 39–53.

Hammerson, Geoffrey A. *Amphibians and Reptiles in Colorado.* Denver: Colorado Division of Wildlife, 1986.

James, Edwin. "Catalogue of Plants Collected During a Journey to and from the Rocky Mountains, During the Summer of 1820." *Transactions of the American Philosophical Society,* n.s., 2 (1825): 172–190.

———, ed. *Account of an Expedition from Pittsburgh to the Rocky Mountains, Performed in the Years 1819 and '20, by Order of the Hon. J.C. Calhoun, Sec. of War: Under the Command of Major Stephen H. Long* (1823). In Reuben Gold Thwaites, vols. 14–17. *Early Western Travels, 1748–1846,* edited by Cleveland: Clark, 1905. [Also available in Readex Microprint, 2 Vols., 1966.]

Lecompte, Janet. "James Pursley." In *The Mountain Men and the Fur Traders of the Far West,* edited by L. R. Hafen, 8: 277–285. Glendale, Calif.: Clark, 1971.

BIBLIOGRAPHY

Mawdsley, J. R. "The Entomological Collections of Thomas Say." *Psyche* 100 (1993): 163–171.

McDermott, J. F. "Samuel Seymour: Pioneer Artist of the Plains and Rockies." *Annual Report of the Smithsonian Institution,* 1950, 497–509.

McKelvey, Susan D. *Botanical Explorations of the Trans-Mississippi West, 1790–1850.* Jamaica Plain, Mass.: Arnold Arboretum, Harvard University, 1955.

McLarty, V. K, ed. "The First Steamboats on the Missouri: Reminiscences of Captain W. D. Hubbell." *Missouri Historical Review* 51 (1957): 373–381.

Murphy, Robert C. "The sketches of Titian Ramsay Peale, 1799–1855." *Proceedings of the American Philosophical Society* 101 (1957): 523–531.

Nichols, Roger L., and Patrick L. Halley. *Stephen Long and American Frontier Exploration.* Newark: University of Delaware Press, 1980.

Nuttall, Thomas. *Journal of Travels into the Arkansa Territory, During the Year 1819, with Occasional Observations on the Manners of the Aborigines* (1821). In *Early Western Travels, 1748–1846,* edited by Reuben Gold Thwaites, vol. 13. Cleveland: Clark, 1905.

Oglesby, Richard E. *Manuel Lisa and the Opening of the Missouri Fur Trade.* Norman: University of Oklahoma Press, 1963.

Ord, George. 1859. "A Memoir of Thomas Say." In *The Complete Writings of Thomas Say on the Entomology of North America,* edited by John L. LeConte, vii–xxi. New York: Bailliere Brothers, 1859.

Osterhout, G. E. "Concerning the Ornithology of the Long Expedition of 1820". *Oologist* 37 (1920): 118–120.

———. "Rocky Mountain Botany and the Long Expedition of 1820". *Bulletin of the Torrey Botanical Club* 47 (1920): 555–562.

Peale, Titian Ramsey. "Journal of Titian Ramsay Peale, Pioneer Naturalist." Edited by A. O. Weese. *Missouri Historical Review* 41 (1947): 147–163.

Poesch, Jessie. *Titian Ramsay Peale, 1799–1885, and His Journals of the Wilkes Expedition.* Philadelphia: American Philosophical Society, 1961.

Porter, Charlotte M. "The Life Work of Titian Ramsay Peale." *Proceedings of the American Philosophical Society,* 129 (1985): 300–312.

Say, Thomas. *The Complete Writings of Thomas Say on the Entomology of North America.* Edited by John L. LeConte. 2 vols. New York: Bailliere Brothers,1859.

Stroud, Patricia Tyson. *Thomas Say: New World Naturalist*. Philadelphia: University of Pennsylvania Press, 1992.

Terrell, John Upton. *Zebulon Pike: The Life and Times of an Adventurer*. New York: Weybright and Talley, 1968.

Torrey, John. Descriptions of Some New or Rare Plants from the Rocky Mountains collected in July 1820 by Dr. Edwin James. *Annals of the New York Lyceum of Natural History*, 1(1824) : 30–36, 148–156; 2 (1827): 161–254.

Torrey, John, and Asa Gray. *A Flora of North America*. 2 vols. 1838, 1843. Reprint. New York: Hafner, 1969.

Townsend, John Kirk. *Across the Rockies to the Columbia*. 1838. Reprint. Lincoln: University of Nebraska Press, 1978.

Trenton, Patricia, and P. H. Hassrick. *The Rocky Mountains: A Vision for Artists in the Nineteenth Century*. Norman: University of Oklahoma Press, 1983.

Tucker, John M. "Major Long's Route from the Arkansas to the Canadian River, 1820." *New Mexico Historical Review* 38 (1963): 185–219.

Tyler, Ron. *Prints of the West*. Golden, Colo.: Fulcrum, 1994.

Viola, Herman J. *Exploring the West*. Washington, D. C.: Smithsonian Institution Press, 1987.

Voelker, Frederic E. "Ezekiel Williams." In *The Mountain Men and the Fur Trade of the Far West*, edited by L. R. Hafen, 9: 393–409. Glendale, Calif.: Clark, 1972.

Webb, Walter Prescott. "The American West: Perpetual Mirage." *Harper's Magazine*, May 1957, 26–31.

Weber, W. A., and R. C. Wittman. *Catalog of the Colorado Flora: A Biodiversity Baseline*. Niwot: University Press of Colorado, 1992.

Weiss, H. B., and G. M. Zeigler. *Thomas Say, Early American Naturalist*. Springfield, Ill.: Thomas, 1931.

Wood, Richard G. *Stephen Harriman Long, 1784–1864: Army Engineer, Explorer, Inventor*. Glendale, Calif.: Clark, 1966.

Worster, Donald. *Under Western Skies: Nature and History in the American West*. New York: Oxford University Press, 1992.

Zwinger, Ann H., and Beatrice E. Willard. *Land Above the Trees: A Guide to American Alpine Tundra*. New York: Harper & Row, 1972.

INDEX

258

INDEX